EXPERIMENTS
IN
PHYSICAL CHEMISTRY

EXPERIMENTS IN PHYSICAL CHEMISTRY

Second Revised and Enlarged Edition

J. M. WILSON,
M.Sc., F.R.I.C.
Senior Lecturer in Physical Chemistry, University of Technology, Loughborough

R. J. NEWCOMBE,
M.Sc., F.R.I.C., A.Inst.P.
Head of Department of Science and Mathematics, College of Technology, Grimsby

A. R. DENARO,
M.Sc., Ph.D., F.R.I.C.
Principal Lecturer in Physical Chemistry, College of Technology, Liverpool

R. M. W. RICKETT,
Ph.D., F.R.I.C.
Vice-Principal, Sir John Cass College, London

THE QUEEN'S AWARD
TO INDUSTRY 1966

PERGAMON PRESS

OXFORD · LONDON · EDINBURGH · NEW YORK
TORONTO · SYDNEY · PARIS · BRAUNSCHWEIG

Pergamon Press Ltd., Headington Hill Hall, Oxford
4 & 5 Fitzroy Square, London W.1
Pergamon Press (Scotland) Ltd., 2 & 3 Teviot Place, Edinburgh 1
Pergamon Press Inc., 44–01 21st Street, Long Island City, New York 11101
Pergamon of Canada Ltd., 207 Queen's Quay West, Toronto 1
Pergamon Press (Aust.) Pty. Ltd., 19a Boundary Street,
Rushcutters Bay, N.S.W. 2011, Australia
Pergamon Press S.A.R.L., 24 rue des Écoles, Paris 5ᵉ
Vieweg & Sohn GmbH, Burgplatz 1, Braunschweig

First edition 1962
Second Revised and enlarged edition 1968

Library of Congress Catalog Card No. 68-18536

PRINTED IN GREAT BRITAIN BY J. W. ARROWSMITH LTD., BRISTOL

08 012541 7

Contents

Preface

Experiment is the interpreter of nature. Experiments never deceive. It is our judgement which sometimes deceives itself because it expects results which experiment refuses.

LEONARDO DA VINCI

THIS book is designed for students reading chemistry for the degree of B.Sc. Hons. of a British University or of the Council for National Academic Awards and also for the Graduate membership examination of the Royal Institute of Chemistry. It would also be suitable for students in the Junior and Senior Years of an American University who are majoring in Chemistry.

In each experiment there is a discussion of the theory associated with the objective involved. This theoretical discussion does not claim to be exhaustive but we hope that it is adequate to illustrate the point of the experiment and that it will provide a link between the theory and practice of physical chemistry. In every case the student should consult other textbooks, especially the theoretical books, in order to supplement the information given.

A course of practical physical chemistry which consists of about a hundred experiments cannot claim to be entirely original. Some of the standard experiments which are to be found in other books will also be found here. In this respect we are indebted to the authors of other textbooks and by way of acknowledgement those books to which we have had the greatest recourse are listed in Appendix II.

The book is divided into three parts. It is difficult to make rigorous divisions but we have been guided by the following principles. Part I consists of those experiments which, in general, have a simple theoretical background. Part II consists of experiments which are associated with more advanced theory or more recently developed techniques, or which require a greater degree of experimental skill. Part III contains experiments which are in the nature of investigations. These investigations may be regarded as minor research projects which are suitable for final-year students. The instructions given in these cases are intended to serve only as a framework and the investigations may be elaborated as time permits.

In those experiments which require some degree of instrumentation it will be found that particular instruments have been listed. We do not suggest that these instruments are the only ones suitable for a particular experiment but we mention them because our students have found them to be satisfactory for the task assigned.

In short, this book has been written to facilitate experimental work in the physical chemistry laboratory at every stage of a student's career. It is hoped that by following this course the student will gain confidence in his ability to perform a physical chemistry experiment and to appreciate the value of the experimental approach.

We should like to express our thanks to Professor K. S. W. Sing (formerly Head of the Department of Chemistry, Liverpool), and to Professor R. F. Phillips (Head of the Department of Chemistry, Loughborough University of Technology) in whose laboratories the greater part of the experimental work was done. We are indebted to a number of our colleagues mentioned in Part III and the Appendixes for the contributions given with their names. We are also indebted to other colleagues for contributions to some of the experiments or investigations: Dr. J. Cast (Exps. 65 and 66), Dr. G. G. Jayson (Exps. 88 and 89), Dr. A. L. Smith (Exps. 6 and 51), Dr. M. W. T. Pratt (Exp. 32), Dr. G. G. J. Boswell (Exp. 63), Mr. N. Gore (Inv. 101) and Mr. P. Tickle (Inv. 112). We thank Dr. O. E. Ford, Dr. R. Stock and other colleagues for useful discussion, and our students who tested many of the new and modified experiments. In conclusion our gratitude is due to our wives for their forbearance and assistance during the course of the preparation of this work.

J. M. W.   R. J. N.

A. R. D.   R. M. W. R.

List of Symbols

EXCEPT in a few cases, the symbols used are those recommended in the Report by the Symbols Committee of the Royal Society, representing the Royal Society, the Chemical Society, the Faraday Society, and the Physical Society, issued in 1951, and amended by that Committee in 1959.

A	ampère.
A	absorbance.
Å	ångström unit.
a	activity, area, van der Waals' constant.
b	van der Waals' constant.
°C	degree Centigrade.
C	capacitance.
	with subscript: molar heat capacity.
c	concentration.
	with subscript: specific heat capacity.
D	debye.
d	density.
∂	partial differential.
E	energy, electromotive force of voltaic cell, potential of an electrode.
e	base of natural logarithm.
e	proton charge, electron charge.
F	Faraday.
f	activity coefficient.
G	Gibbs function.
g	gram.
g	acceleration of gravity, osmotic coefficient, gaseous state.
H	enthalpy.
h	Planck's constant, height.
I	ionic strength, intensity of light.
K	Kelvin, chemical equilibrium constant, molal b.p. elevation constant, molal f.p. depression constant.
k	Boltzmann's constant, velocity constant of chemical reaction, constant.
l	length, liquid state.
l.	litre.
M	molar concentration.
M	molecular weight.
m	mass, molality.
N	normal concentration.
N	Avogadro's number.
n	number of moles, transport number, refractive index.
P	pressure, polarization.
p	pressure.
R	gas constant, electrical resistance, refraction.

r	radius.
S	entropy.
s	solubility, solid state.
T	temperature on Kelvin scale, transmittance.
t	time, temperature not on Kelvin scale.
U	internal energy.
V	volt.
V	molar volume.
v	volume.
w	weight.
x	mole fraction.
z	valency of an ion, number of electrons in an electrode reaction.
α	degree of dissociation, angle of optical rotation.
Γ	surface concentration.
γ	ratio of heat capacities, surface tension, activity coefficient, Raoult law factor.
Δ	increment.
ϵ	electric permittivity, molar absorptivity.
η	viscosity.
Λ	equivalent conductance.
λ	wavelength, equivalent ionic conductance at infinite dilution.
μ	micron, chemical potential, dipole moment, Joule–Thomson coefficient.
ν	frequency.
π	ratio of circumference to diameter.
K	conductance.
κ	specific conductance, quantity proportional to the square root of ionic strength and having the dimensions of reciprocal length.

ln, log   logarithms to bases e and 10, respectively.

PART I

Molecular Weight using van der Waals' Equation

Discussion

For one mole of gas, van der Waals' equation of state may be written

$$(P + a/V^2)(V - b) = RT$$

where a and b are constants. Hence the molar volume V is given by the cubic equation

$$PV - Pb + a/V - ab/V^2 = RT$$

Substituting P/RT for $1/V$ in the third term and neglecting the small term ab/V^2 the equation can be written

$$PV = RT - P\left(\frac{a}{RT} - b\right) \qquad (1)$$

If the constants a and b are known the volume of one mole of gas at the temperature and pressure of the experiment can be calculated.

The weight, w grams, of a definite volume of nitrogen or carbon dioxide, say v litres, may be determined by experiment. As

$$M/w = V/v \qquad (2)$$

the molecular weight M of the gas can be calculated.

Apparatus and Chemicals

Glass globe, wooden box, wide mouth bottle, rubber stopper, three stop cocks, silica gel tower, T-shaped glass tap, mercury manometer, vacuum pump, cylinder of nitrogen or of carbon dioxide, and lute.

Method

The glass globe is evacuated, weighed and then filled with boiled distilled water. The globe and contents are immersed in a water bath and the system is allowed to come to thermal equilibrium. During this period the tap of the globe should be left open. The weight of the globe full of water is determined. Knowing the density of the water at the temperature of the experiment, the volume of the globe can be calculated.

The globe is emptied, cleaned with chromic acid mixture, rinsed with water, and dried using alcohol and ether. The key of the tap is adequately greased. The globe is evacuated and weighed.

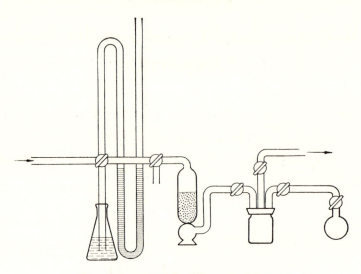

Fig. 1. Apparatus for filling gas globe.

The globe is filled with nitrogen or carbon dioxide to a pressure slightly greater than atmospheric. Whilst the globe is being filled it is kept in the wooden box. When thermal equilibrium is attained the three-way tap is opened and the pressure adjusted to that of the atmosphere, which is read from the barometer. The globe containing the gas is weighed.

The values of the van der Waals' constants a and b for nitrogen and carbon dioxide are

	a	b
nitrogen	1·390 l.^{2} atm mole^{-2}	0·03913 l. mole^{-1}
carbon dioxide	3·592 l.^{2} atm mole^{-2}	0·04267 l. mole^{-1}

These values are to be used when the pressure is in atmospheres and the volume is in litres. The gas constant R is 0·08206 l. atm/mole per °C. Using these values, the volume of one mole of gas at the temperature and pressure of the experiment can be calculated from equation (1).

The molecular weight of the gas can be calculated from equation (2).

The Density of a Liquid as a Function of Temperature

Discussion

The density of a liquid is the mass of unit volume of the liquid. The generally accepted unit of volume is the millilitre which is defined as the volume occupied by 1 g of water at the temperature of maximum density (4°C). The density of a liquid at $t°C$ is expressed relative to that of water at 4°C and is represented by the symbol

$$d_{4°(H_2O)}^{t°(liq.)}$$

The density of a liquid at a particular temperature t is the product of the relative density of the liquid

$$d_{t°(H_2O)}^{t°(liq.)}$$

(the ratio of the weight of a given volume of the liquid to the weight of the same volume of water at the same temperature) and the density of water at that temperature, i.e.

$$d_{4°(H_2O)}^{t°(liq.)} = d_{t°(H_2O)}^{t°(liq.)} \times d_{4°(H\ O)}^{t°(H_2O)}$$

$$= \frac{w'}{w} \times d_{4°(H_2O)}^{t°(H_2O)}$$

where w is the apparent weight of water and w' the apparent weight of liquid at the temperature t. In order to arrive at an accurate result all weighings must be corrected for the buoyancy of air, i.e.

$$d_{4°(H_2O)}^{t°(liq.)} = \frac{w'}{w} \times d_{4\ (H_2O)}^{t°(H_2O)} - \frac{0·0012(w'-w)}{w}$$

where the factor 0·0012 is the mean density of air.

The volume V_t of m grams of a liquid at $t°C$ is related to the volume V_0 of the same mass at 0°C by the equation

$$V_t = V_0(1+\alpha t)$$

3

As

$$V_t/m = (V_0/m)(1+\alpha t)$$

then

$$1/d_t = (1/d_0)(1+\alpha t)$$

A plot of $(1/d_t)$ against t will have a slope of (α/d_0) and an intercept, when $t = 0$, of $(1/d_0)$. Thus, α, the coefficient of expansion, may be determined.

The internal pressure P_i of a liquid is a measure of molecular attractive forces. It is defined by

$$P_i = (\partial U/\partial V)_T$$

where U is the internal energy. It can be shown that $P_i = T(\alpha/\beta)$ where β is the coefficient of compressibility.

Apparatus and Chemicals

Two density bottles or two pyknometers, pyknometer support, pyknometer caps, cyclohexane, benzene and a thermostat at 25°C.

Method

The pyknometer and caps are cleaned and dried thoroughly and then weighed. The pyknometer is filled with distilled water by attaching a rubber tube to one end and sucking gently whilst the other end of the pyknometer is immersed in the water.

The pyknometer is suspended in a thermostat in such a way that only the arms of the pyknometer are above the surface of the thermostat liquid.

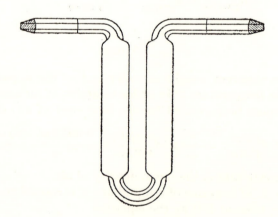

FIG. 1. A pyknometer.

After allowing 15–20 min for thermal equilibrium to be established, the amount of water in the pyknometer is adjusted so that the pyknometer is filled from the tip of the jet of one limb up to the mark on the other limb. Water may be added by placing a glass rod carrying a drop of water in contact with the jet of the limb which is already full and water may be withdrawn by placing a filter paper in contact with the same jet. A short time should be allowed for the system to regain thermal equilibrium.

The caps are placed on the pyknometer which is then removed from the thermostat and the outside carefully dried (care is taken not to expel water from the pyknometer by the heat of the hand). The pyknometer and contents are weighed after allowing a short time for them to take the temperature of the balance case.

The pyknometer and caps are emptied and dried and the above procedure repeated with the apparatus filled with the liquid under investigation.

If density bottles are used instead of pyknometers, the stoppers of the bottles should only be inserted when the contents have reached the required temperature. Density bottles are easier to use than pyknometers and for most purposes they are sufficiently accurate.

Values of the density of water at various temperatures are obtained from tables and the densities of benzene and cyclohexane are determined at four different temperatures. A graph of $(1/d_t)$ against t is plotted and the coefficient of expansion, α, is calculated. The internal pressure of each liquid is also calculated given that β has the values $1{\cdot}08 \times 10^4$ atm^{-1} for cyclohexane and $0{\cdot}94 \times 10^4$ atm^{-1} for benzene at 20°C.

Molar Refraction

Discussion

The specific refraction of a pure liquid is given by the formula

$$R_s = \frac{n^2 - 1}{n^2 + 2} \cdot \frac{1}{d}$$

where n is the refractive index and d the density. The molar refraction R_m is given by the formula

$$R_m = R_s \cdot M$$

where M is the molecular weight.

Molar refraction is an additive property, therefore, if it is determined for different members of an homologous series the contribution of the —CH_2— group may be obtained.

Apparatus and Chemicals

Abbé refractometer, semi-micropipette, cotton wool, ethyl acetate, propyl acetate and butyl acetate.

Method

The mirror of the refractometer may be illuminated with the light from a pearl lamp. Measurements may be made at room temperature. (Note this temperature for each determination.) No water should be circulated through the prism for this experiment.

A few drops of the liquid are placed on the ground-glass surface of the lower prism of the refractometer using the semi-micropipette. (Caution: do not scratch the surface, and when subsequently cleaning the prism, use a piece of cotton wool.) The prism box is closed and clamped. Volatile liquids should be introduced through the groove of the prism box. The telescope is focused upon the cross hairs in the field by turning the eyepiece. The edge of the dark band is brought to the intersection of the cross hairs and the compensator adjusted until the coloured fringe disappears and a sharp dark-white contrast is obtained. The instrument is carefully adjusted until the intersection of the cross hairs and the dark edge of the band coincide. The refractive index is read on the scale with the aid of the reading telescope.

The refractive indices of ethyl acetate, propyl acetate and butyl acetate are determined and the results recorded in tabular form under the headings:

$$\text{Ester} \quad n \quad t°C \quad d \quad R_s \quad M \quad R_m$$

The densities at $t°C$ for ethyl acetate, propyl acetate and butyl acetate are obtained from tables.

From the results the contribution to the molar refraction made by the $—CH_2—$ group may be calculated.

The experiment is repeated with a second homologous series such as ethanol, propanol and butanol or with hexane, cyclohexane and benzene. The results are discussed in terms of the additive concept of molecular refraction.

Viscosity as a Function of Temperature

Discussion

Viscosity is the resistance that one part of a fluid offers to the flow of a contiguous part of the fluid. The coefficient of viscosity η is the force required per unit area to maintain unit difference of velocity between two parallel planes in the fluid, 1 cm apart. The unit of viscosity in c.g.s. units is called a poise.

The viscosity of a liquid may be determined by measuring its rate of flow through a capillary tube. For a liquid flowing through a capillary tube of radius r, for a time t, under a constant pressure head p, the volume v of liquid issuing from the tube is given by Poiseuille's equation

$$v = \frac{\pi p t r^4}{8l\eta} \tag{1}$$

where l is the length of the tube. If the dimensions of the capillary and the volume of the liquid flowing through it are constant, equation (1) reduces to

$$\eta = kpt \tag{2}$$

Thus although the determination of absolute viscosity is a matter of some difficulty, the ratio of the viscosities of two liquids may be readily determined using a viscometer. The pressure p at any instant driving the liquid of density d through the capillary of a viscometer is hdg, where h is the difference in height between the levels in each limb of the instrument. Although h varies throughout the experiment the initial and final values are the same in every case; hence p is proportional to the density. The relationship between the viscosities η_1 and η_2 of two liquids 1 and 2 having densities d_1 and d_2 is

$$\frac{\eta_1}{\eta_2} = \frac{d_1 t_1}{d_2 t_2} \tag{3}$$

where t_1 and t_2 are the times of flow. The viscometer must therefore be calibrated by using a liquid of known viscosity and density, e.g. water.

The expansion of the glass with temperature modifies both k and p and the time of flow depends upon $1/kp$. The viscometer must be calibrated with water at each temperature.

The variation of the viscosity of a liquid with temperature may be expressed by the equation

$$\eta = A \cdot e^{E/RT}$$

i.e.

$$\ln \eta = \frac{E}{RT} + \ln A$$

where A is a constant and E is a measure of the energy required to overcome the elementary flow process.

Apparatus and Chemicals

Ostwald viscometer, stop watch, 10 ml graduated pipette, thermostats at 25°C, 30°C and 40°C, rubber tubing and benzene.

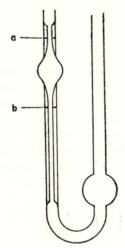

FIG. 1. An Ostwald viscometer.

Method

The viscometer is cleaned with chromic–sulphuric acid mixture and rinsed with distilled water and dried. The viscometer is clamped in the thermostat at 25°C. An exactly specified volume of water is added from the pipette and 10 min is allowed for the water to reach the temperature of the thermostat. With the aid of a piece of rubber tubing the liquid is sucked up the capillary arm of the viscometer until the surface of the liquid is above the upper mark a (Fig. 1). The liquid is then allowed to flow down the arm. The time required for the surface of the liquid to pass from the upper mark a to the lower mark b is noted.

The process is repeated using the same volume of benzene. The whole procedure is then repeated at 30°C and 40°C. A graph is then plotted of log η vs. $1/T$. This graph should be a straight line of slope $E/2.303R$ from which E may be determined.

The Ratio of the Heat Capacities of a Gas
(Clément and Desormes Method)

Discussion

In this method a large vessel is filled with air at a pressure very slightly greater than atmospheric pressure. The system is opened so that the gas expands rapidly to atmospheric pressure and the system is closed again. The expansion is so rapid that it is virtually adiabatic and the gas cools in the process. As the gas warms up to the original temperature, the pressure in the system increases until once again it is greater than atmospheric. From measurements of the initial and final pressures, the ratio of the heat capacities of air γ may be calculated.

Let the initial pressure of the gas be p_1 and the final pressure p_2. Let the atmospheric pressure be p, and let the original and final temperature of the gas be T. Suppose that the intermediate temperature to which the gas cools after the adiabatic expansion is T'.

For the adiabatic expansion we may write

$$\left(\frac{p_1}{p}\right) = \left(\frac{T}{T'}\right)^{\gamma/(\gamma-1)} \tag{1}$$

and for the final change

$$\frac{p}{p_2} = \frac{T'}{T} \tag{2}$$

Substituting in equation (1) we have

$$\left(\frac{p_1}{p}\right) = \left(\frac{p_2}{p}\right)^{\gamma/(\gamma-1)}$$

or

$$(\gamma-1)\ln\frac{p_1}{p} = \gamma\ln\frac{p_2}{p} \tag{3}$$

Rearranging equation (3)

$$\gamma = \frac{\ln p_1 - \ln p}{\ln p_1 - \ln p_2}$$

If the *excess* pressures are small then

$$\gamma = \frac{p_1 - p}{p_1 - p_2}$$

The initial excess pressure $(p_1 - p)$ is given by the initial difference of the levels h_1, of the oil in the manometer (see Fig. 1), and the final excess pressure $(p_2 - p)$ is given by the final difference of the levels h_2 of the manometer oil. Hence

$$p_1 - p_2 = h_1 - h_2$$

and

$$\gamma = \frac{h_1}{h_1 - h_2} \tag{4}$$

Apparatus and Chemicals

Carboy, oil manometer, squeeze bulb for pumping air into the carboy.

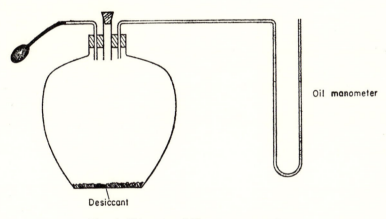

Oil manometer

Desiccant

Fig. 1. Clément and Desormes apparatus.

Method

Air is pumped into the carboy until the manometer registers an excess pressure of about 20 cm of oil. A screw clip is applied to the connection through which air has been admitted and the reading on the manometer is noted when it has become steady.

The bung is withdrawn from the wide bore tube carried in the carboy stopper so that the air inside the carboy may expand to atmospheric pressure. The bung is then rapidly replaced. It is inadvisable to wait for the manometer oil to level before replacing the bung as the oil will probably be too viscous to follow the pressure change closely.

The bung will have to be replaced almost immediately after it has been removed.

When the manometer reading has again become steady the excess pressure in the system is noted.

From the results the ratio of the heat capacities of air may be calculated.

Further experiments should be carried out at successively lower initial pressures to check the results.

EXPERIMENT 6

The Ratio of the Heat Capacities of a Gas (Lummer and Pringsheim Method)

Discussion

The main errors involved in the Clément and Desormes method of determining the ratio of the heat capacities of a gas arise from the rapid heat exchange between the gas and the walls of the vessel during expansion and the difficulty of closing the outlet at the exact moment when the external and internal pressures are equal. For these reasons the calculated value of γ is always too low.

An improved method is due to Lummer and Pringsheim who measured the temperature of the gas at the centre of the vessel by means of a platinum bolometer before and after expansion. The value of γ may thus be obtained directly from equation (1) in Experiment 5.

$$\frac{p_1}{p} = \left(\frac{T}{T'}\right)^{\gamma/(\gamma-1)} \tag{1}$$

Hence

$$(\gamma-1)\ln\frac{p_1}{p} = \gamma\ln\frac{T}{T'}$$

and

$$\gamma = \frac{\ln(p_1/p)}{\ln(p_1/p) - \ln(T/T')} \tag{2}$$

In this variation the external (atmospheric) pressure must be known but one error is avoided since the outlet is left open.

In this experiment the temperature of the gas is measured with a calibrated thermistor. The resistance of a thermistor varies with temperature according to the equation

$$R = Ae^{b/T}$$

where A and b are constants which may be determined from calibration experiments. The resistance of the thermistor is measured with a Wheatstone bridge circuit.

13

For the vessel provided in Experiment 5 the "time constant" of the temperature recovery after expansion is probably about 20 sec, so that any temperature-measuring device used must have a considerably shorter "time constant". The recommended thermistor has a thermal "time constant" of about 1 sec. It is evident that this thermistor must have very fine connecting wires of appreciable resistance which themselves must have a temperature coefficient of resistance. To compensate for this effect a pair of dummy leads are provided which are joined at their lower ends and which may be included in the Wheatstone bridge circuit. They are physically located adjacent to the thermistor leads.

A high resistance is included in the supply to the bridge circuit to limit the current and thus limit self heating of the thermistor. The complete Wheatstone bridge circuit is illustrated in Fig. 1.

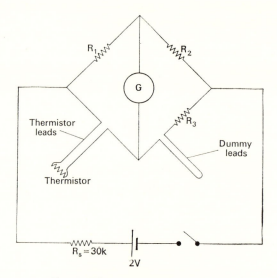

Fig. 1. Wheatstone bridge circuit.

Apparatus and Chemicals

As in Experiment 5 with the addition of calibrated thermistor (S.T.C. Type U 23 US), resistance boxes, galvanometer, accumulator, 30 kilohm resistor.

Method

The vessel is pumped up to a pressure of about 10 cm of manometer oil. The gas is allowed to come to thermal equilibrium as indicated by the constancy of the thermistor resistance.

With the bridge exactly balanced and the resistance of the thermistor noted the bung is withdrawn from the tube carried in the carboy stopper and the maximum galvanometer reading is noted. (It would probably take too long to balance the bridge again.) If the temperature recovery makes it necessary, an extrapolation method must be used to obtain the galvanometer reading immediately after expansion.

A variable resistor is substituted for the thermistor in the bridge circuit and the value of the resistance giving the same deflection on the galvanometer as was observed in the main experiment is determined. This corresponds to the resistance of the thermistor immediately after expansion of the gas. The values of the constants A and b obtained from a calibration of the thermistor are used to convert the thermistor resistances into terms of temperature. The atmospheric pressure is read from a barometer and γ for the gas is calculated from equation (2).

Simplify equation (2) in a similar manner to that used in Experiment 5. Is the simplification justified in this case?

Molecular Weight by Ebullioscopy (Landsberger's Method)

Discussion

If a solution of a non-volatile solute is sufficiently dilute it will obey Raoult's law and the mole fraction of the solvent x_1 in the solution is related to the molar heat of vaporization ΔH_v of the solvent at the boiling point T of the solution by

$$\left[\frac{\partial \ln x_1}{\partial T}\right]_p = -\frac{\Delta H_v}{RT^2} \tag{1}$$

Assuming that the heat of vaporization is independent of temperature, equation (1) may be easily integrated and applying the condition that when $x_1 = 1$, $T = T_0$, the boiling point of the pure solvent, the result is given by

$$\ln x_1 = -\frac{\Delta H_v}{R}\left(\frac{T-T_0}{TT_0}\right) \tag{2}$$

Putting $(T-T_0) = \Delta T_v$ and making the approximation that $TT_0 \approx T_0^2$ equation (2) becomes

$$\ln x_1 = -\frac{\Delta H_v}{RT_0^2}\Delta T_v \tag{3}$$

Now $\ln x_1 = \ln(1-x_2) \approx -x_2$, where x_2 is the mole fraction of solute in the solution. Furthermore, as the solution is dilute we may make the approximation $x_2 \approx n_2/n_1$, where n_2 and n_1 are the number of moles of solute and solvent, respectively, in the solution. Equation (3) may thus be written

$$\Delta T_v = K_v m$$

where m is the molality of the solution and K_v is given by

$$K_v = \frac{RT_0^2}{\Delta H_v n}$$

n being the number of moles of solvent contained in 1000 g of solvent. K_v is known as the molal boiling point elevation.

16

If w_2 g of the solute of molecular weight M_2 are dissolved in w_1 g of the solvent

$$M_2 = \frac{1000K_v w_2}{w_1 \Delta T_v}$$

A determination of the elevation of the boiling point of a solvent caused by the addition of a solute to give a dilute solution may therefore be used in the calculation of the molecular weight of the solute.

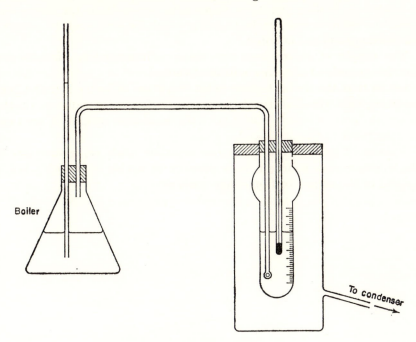

FIG. 1. Landsberger apparatus.

The method used in this experiment is due originally to Landsberger. The solution is heated to boiling point by passing in the vapour of the solvent. In this way it is possible to avoid superheating.

Apparatus and Chemicals

Landsberger apparatus, thermometer graduated in 0·1°C and urea.

Method

The graduated tube is thoroughly cleaned by washing it with distilled water. About 5 ml of distilled water is placed in the graduated tube, the boiler is half filled with water and the apparatus assembled. The water in the boiler is heated. Eventually steam will pass into the graduated tube and the contents will be heated to boiling point. When thermal equilibrium has been established the reading on the thermometer is noted. By this time the volume of water in the graduated tube will have increased due to condensation of steam from the boiler.

Most of the contents are emptied from the tube until only about 5 ml remain. Into this remaining water about 1·0 g of urea is accurately weighed and the apparatus is assembled again. The solution of urea is heated to boiling point by passing in steam from the boiler and when the solution has reached boiling point the temperature recorded on the thermometer is noted.

The supply of steam from the boiler is immediately disconnected and the volume of the urea solution is noted.

Given that the molal boiling point elevation of water is 0·51°C and assuming that the density of the urea solution is the same as that of water at 100°C which is 0·96 g ml^{-1} the molecular weight of urea can be calculated.

A series of three or four determinations are made using successively larger volumes of solution. Is it possible to decide if Raoult's law is obeyed in the concentration range studied?

Molecular Weight by Ebullioscopy
(Cottrell's Method)

Discussion

The theory of the ebullioscopic technique has already been discussed in connection with the Landsberger method in Experiment 7 and reference should be made to this discussion.

In the Landsberger method the temperature of the boiling solution is measured with a thermometer, the bulb of which is immersed in the solution. This may lead to a slight inaccuracy as the temperature of a boiling liquid a few centimetres below the surface will be higher than the surface temperature where the liquid is in equilibrium with the vapour. Cottrell's method avoids this difficulty by locating the bulb of the thermometer in the vapour and pumping the boiling liquid so that it cascades over the thermometer bulb. In this way the thermometer records the temperature of the boiling liquid when it is in equilibrium with its vapour.

Apparatus and Chemicals

Cottrell's apparatus, condenser, Beckmann thermometer, microburner, chloroform and naphthalene.

Method

A weighed quantity of chloroform, sufficient to cover the lower portion of the pump as illustrated in Fig. 1, is added to the apparatus. A small flame from the microburner is applied to the wire sealed into the bottom of the apparatus until the boiling liquid is being pumped over the bulb of the Beckmann thermometer. Once pumping is proceeding regularly the flame is not altered. The reading on the Beckmann thermometer should be near the bottom of the scale. If this is not so the thermometer should be adjusted according to the instructions issued by the manufacturers. The boiling point of the pure solvent as recorded by the thermometer is noted.

In order to determine the molal boiling point elevation, a known weight of solute of known molecular weight, e.g. naphthalene, is added. The amount of solute added should be sufficient to produce an

19

elevation of about 0·2°C. The boiling point of the solution is now determined.

Two further quantities of naphthalene are added, the boiling point of each solution being determined. A plot of the boiling points of the solutions against the weight of naphthalene added should be a straight line.

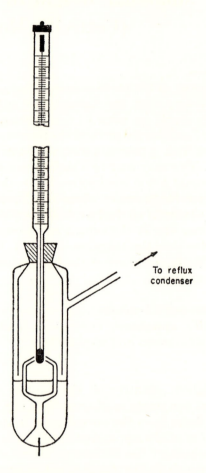

To reflux condenser

FIG. 1. Cottrell's apparatus.

The molal boiling point elevation of the solvent is calculated and the experiments are repeated with an unknown solid to determine its molecular weight.

Molecular Weight by Cryoscopy

Discussion

If a solution is sufficiently dilute the solvent will obey Raoult's law. Under these circumstances, and if at the freezing point of the solution it is in equilibrium with pure solid solvent, the mole fraction of solvent x_1 in the solution is related to the molar heat of fusion ΔH_f of the solvent at the freezing point T of the solution by

$$\left[\frac{\partial \ln x_1}{\partial T}\right]_p = \frac{\Delta H_f}{RT^2} \tag{1}$$

Assuming that the heat of fusion is independent of temperature equation (1) may be easily integrated and applying the condition that when $x_1 = 1$, $T = T_0$, the freezing point of the pure solvent, the result is given by

$$\ln x_1 = \frac{\Delta H_f}{R}\left(\frac{T - T_0}{TT_0}\right) \tag{2}$$

Putting $(T_0 - T) = \Delta T_f$ and making the approximation that $TT_0 \simeq T_0^2$, equation (2) becomes

$$\ln x_1 = -\frac{\Delta H_f}{RT_0^2}\Delta T_f \tag{3}$$

Now $\ln x_1 = \ln(1 - x_2) \simeq -x_2$, where x_2 is the mole fraction of solute in the solution. Furthermore, as the solution is dilute we may make the approximation $x_2 \simeq n_2/n_1$, where n_2 and n_1 are the number of moles of solute and solvent respectively in the solution. Equation (3) may thus be written

$$\Delta T_f = K_f m$$

where m is the molality of the solution and K_f is the molal freezing point depression given by

$$K_f = \frac{RT_0^2}{\Delta H_f n}$$

n being the number of moles of solvent contained in 1000 g of solvent.

21

Apparatus and Chemicals

Beckmann freezing point apparatus, Beckmann thermometer and benzene.

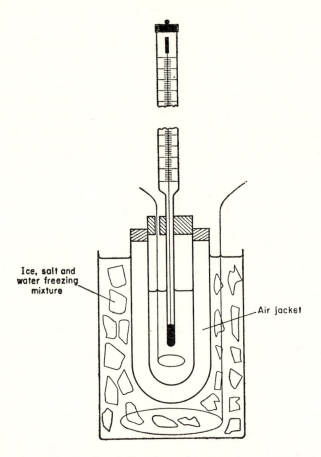

Fig. 1. Freezing point apparatus.

Method

The Beckmann thermometer should be adjusted in accordance with the manufacturers' instructions so that a reading near the top of the scale corresponds to about 5°C.

A known weight of benzene (about 20 g) is added to the centre tube of the apparatus. The thermometer and stirrer are fitted and the tube is cooled by direct immersion in the ice bath until the benzene starts to freeze. At this point the tube is warmed slightly and then set in the jacket in the ice bath. The benzene is stirred continuously and the falling temperature closely watched. Generally supercooling occurs; hence when the solid separates on freezing, the temperature rises to the true freezing point. This temperature is noted. The system should then be warmed and the operation repeated in order to re-determine the freezing point.

The tube is removed, the benzene melted with gentle warming and a known weight of an unknown solid is added to the benzene. Sufficient solid should be added to produce a depression of the freezing point of about 0·2°C. The freezing point of this solution is determined and then re-determined as a check. A further quantity of solid is added and the freezing point re-determined; in all, four additions should be made.

A graph of the depression of the freezing point against the weight of substance added is plotted. Given that for benzene:

$$T_0 = 278\cdot5\,°K, \qquad \Delta H_f = 2\cdot382\ \text{kcal mole}^{-1},$$

the molal freezing point depression is calculated and the molecular weight of the unknown solid may be determined.

Molecular Weight by Rast's Method

Discussion

The molal depression constant is characteristic of a solvent and is defined as the lowering of the freezing point caused by dissolving 1 g molecule of solute in 1000 g of solvent.

Camphor possesses a very high molal depression constant, hence if it is used as a solvent the molecular weight of a solute may be determined using an ordinary thermometer and without the use of any special apparatus. The molal depression constant varies considerably with the conditions of the experiment and it should be determined by the use of a compound of known molecular weight (e.g. naphthalene).

Apparatus and Chemicals

Melting point apparatus, thermometer, melting point tubes, naphthalene and camphor.

Method

Some pure camphor is powdered and a small amount transferred to a melting point tube. The melting point tube is placed in the melting point apparatus and heated fairly rapidly until the camphor melts to a clear liquid (about 180°C). The camphor is allowed to cool and solidify. The tube is reheated and carefully observed. The temperature at which the last trace of crystalline material just disappears is noted. This temperature is recorded as the melting point; at least three consistent determinations should be made.

A thin-walled glass tube with a bulbous end is prepared, i.e. a small test tube which has been drawn out in the middle; the tube should be about as large as an ignition tube. The tube is weighed and a few milligrams of naphthalene added and then reweighed. About forty times as much camphor is added and the tube and contents weighed again. The tube is sealed at the constriction. The contents of the tube are subsequently melted by heating gently in an oil bath. To ensure a homogeneous mixture the tube should be rotated repeatedly before cooling.

The melting point of the mixture of naphthalene and camphor is determined; again three consistent determinations are required. From this the molal depression constant is obtained.

A mixture of camphor and the compound of unknown molecular weight is subsequently prepared and the melting point found. Hence from the molal depression constant already calculated, the molecular weight of the unknown compound can be determined.

Heat of Neutralization by Calorimetry

Discussion

In dilute solution strong acids and strong bases may be considered to be completely dissociated into their ions. Furthermore, the salt of a strong acid and a strong base will also be completely dissociated in solution. The neutralization of a strong acid by a strong base may thus be written

$$H^+ + OH^- \rightarrow H_2O$$

The heat effect is therefore independent of the nature of the anion of the acid and the cation of the base.

This is not true if the acid or the base is not completely ionized. Acetic acid is only partially ionized in solution and its neutralization by sodium hydroxide may be written

$$CH_3COOH + OH^- \rightarrow CH_3COO^- + H_2O$$

or, if considered as two stages

$$CH_3COOH \rightarrow CH_3COO^- + H^+$$
$$H^+ + OH^- \rightarrow H_2O$$

The heat of neutralization in this case is the heat of combination of hydrogen ions and hydroxyl ions less the energy which must be used to dissociate any unionized acetic acid molecules.

These heats of neutralization may all be determined by simple calorimetry.

Apparatus and Chemicals

Dewar flask, boiling tube, glass rod, thermometers 0–50°C graduated in 0·1°C, 1 N hydrochloric acid, 1 N sodium hydroxide, 1 N acetic acid and 1 N nitric acid.

Method

The first step is to determine the water equivalent of the apparatus.

Fifty millilitres of distilled water is measured into the Dewar flask with a pipette and the temperature of the water is noted. Let this temperature be t_1°C. Fifty millilitres of distilled water is placed in a flask in a thermostat operating at about 30°C and the flask and contents are allowed to come to thermal equilibrium. The temperature of the water is then noted; let this be $t_2°$C. This water is then poured quickly into the Dewar flask, stirred rapidly and the highest temperature attained is noted. Let this temperature be $t_3°$C.

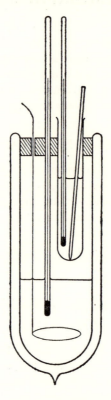

Fig. 1. Calorimeter.

The heat lost by the warm water must be equal to the heat gained by the cold water and the Dewar flask. Hence

$$50(t_2 - t_3) = w(t_3 - t_1) + 50(t_3 - t_1)$$

where w is the water equivalent of the Dewar flask.

The apparatus is now assembled as shown in Fig. 1, with 50 ml of carbonate free normal sodium hydroxide in the Dewar flask and 50 ml of normal hydrochloric acid in the boiling tube carried by the cork in the neck of the Dewar. The solutions are stirred until they are both at the same temperature. Let this temperature be $t_4°$C.

The pointed glass rod is plunged through the bottom of the boiling tube and the two solutions are allowed to mix. The mixture is stirred rapidly and the highest temperature attained is recorded. Let this be $t_5°C$.

Assuming that the total mass of solution is 100 g and that the specific heat of the solution is unity then the heat given out on mixing of the acid and the base is equal to

$$100(t_5 - t_4) + w(t_5 - t_4) = Q \text{ cal (say)}$$

The heat evolved on mixing one litre of normal acid with one litre of normal alkali will therefore be equal to $20Q$ cal. The heat of neutralization of a strong acid and a strong base then, is $20Q$ cal equivalent^{-1}.

If the normalities of the acid and alkaline solutions are not exactly the same, the weaker of the two solutions will govern the extent of the reaction. If x is the normality of the weaker solution and the heat evolved on mixing 50 ml of acid and 50 ml of alkali is Q cal, then the heat of neutralization will be equal to $20Q/x$ cal equivalent^{-1}.

The experiment should be repeated using nitric acid and sodium hydroxide and then using acetic acid and sodium hydroxide.

Heat of Transition by Calorimetry

Discussion

The heat of transition ΔH_t of the reaction

$$\text{Na}_2\text{SO}_4 . 10\text{H}_2\text{O} \rightarrow \text{Na}_2\text{SO}_4 + 10\text{H}_2\text{O}$$

can be calculated if the heats of solution of the two modifications of sodium sulphate are known. If ΔH_a is the heat of solution of the anhydrous salt and ΔH_d the heat of solution of the decahydrate then,

$$\Delta H_t = \Delta H_d - \Delta H_a$$

These heats of solution may be determined by simple calorimetry.

Apparatus and Chemicals

Dewar flask, boiling tube, glass rod, thermometer 0–50°C graduated in 0·1°C, anhydrous sodium sulphate and sodium sulphate decahydrate.

Method

The water equivalent of the apparatus is determined as in Experiment 11.

One hundred millilitres of water is placed in the Dewar flask and allowed to come to a steady temperature. This temperature is noted. In the boiling tube is placed 0·01 mole of the decahydrate in a finely powdered state. The boiling tube is then broken under the water in the Dewar flask, the mixture stirred well, and the maximum temperature attained is noted. From the results ΔH_d is calculated.

Some anhydrous sodium sulphate is dried in an oven and allowed to cool in a desiccator. ΔH_a is determined in a similar manner to that above.

ΔH_t is calculated and the mean temperature at which ΔH_t has been determined is recorded.

Heat of Vaporization by Calorimetry

Discussion

The molar heat of vaporization of a liquid is defined as the quantity of heat required to change one mole of the liquid to the vapour state. A knowledge of the molar heat of vaporization is useful in computing the ebullioscopic constant or the molal boiling point elevation K_v, given by

$$K_v = \frac{RT_0^2}{\Delta H_v n}$$

where the various symbols have the same significance as in Experiment 7.

The molar heat of vaporization is also involved in Trouton's law which may be expressed as

$$\frac{\Delta H_v}{T_0} \backsimeq 21$$

In effect, the law states that the increase in entropy on the vaporization of a liquid at its normal boiling point is approximately 21 cal $°C^{-1}$ mole^{-1}.

Apparatus and Chemicals

Dewar flask, condenser, stop clock, electrical heating unit (30–50 W), ammeter, voltmeter, control resistance, d.c. supply, and carbon tetrachloride.

Method

The Dewar flask is filled about two-thirds full with carbon tetrachloride and the apparatus is assembled as shown in Fig. 1. The electrical heater is switched on and the current is adjusted so that distillation is proceeding smoothly at about one drop per second. When the system has reached thermal equilibrium, i.e. after about 50 ml of distillate has been collected, the receiver is replaced with a weighed flask, and the stop clock is started at the same moment.

The current and the voltage readings must be maintained at a steady value until a further 50 ml of distillate has been collected in the weighed flask. The flask is removed and the clock is stopped at the same time. The flask and contents are weighed.

If the current has been I_1 A, the voltage V_1 V and the duration of the experiment t_1 sec then the electrical energy expended will be $(V_1 I_1 t_1)/4{\cdot}18$ cal. This quantity of heat goes partly to vaporizing the liquid and disappears partly as heat losses from the apparatus. If the molar heat of vaporization of carbon tetrachloride is ΔH_v cal mole^{-1},

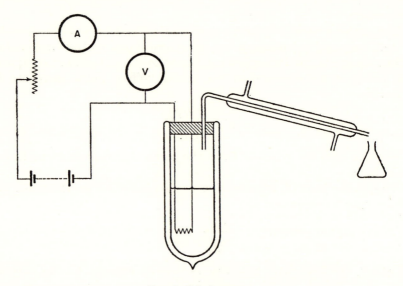

Fɪɢ. 1. Calorimeter.

the weight of distillate w_1 g, the molecular weight of carbon tetrachloride M and the rate of heat loss from the apparatus h cal sec^{-1} then

$$\frac{V_1 I_1 t_1}{4{\cdot}18} = \frac{w_1 \Delta H_v}{M} + h t_1 \tag{1}$$

The experiment is repeated using a current of about $0{\cdot}7$–$0{\cdot}8$ the previous value and a second relationship is obtained

$$\frac{V_2 I_2 t_2}{4{\cdot}18} = \frac{w_2 \Delta H_v}{M} + h t_2 \tag{2}$$

By combination of equations (1) and (2) the unknown heat loss h may be eliminated and ΔH_v may be calculated.

Having calculated the molar heat of vaporization of carbon tetrachloride its molal boiling point elevation is calculated and the constant of Trouton's law is also computed.

The Vapour Pressure of a Liquid as a Function of Temperature

Discussion

The relationship between the vapour pressure of a liquid and the temperature is given by the Clausius–Clapeyron equation

$$\frac{\mathrm{d}\ln p}{\mathrm{d}T} = \frac{\Delta H}{RT^2} \tag{1}$$

Assuming that the heat of vaporization is constant, integration of equation (1) gives

$$\log p = -\frac{\Delta H}{2 \cdot 303 RT} + \text{constant}$$

where ΔH is the heat of vaporization of the liquid.

If the vapour pressure of a liquid is determined at a series of temperatures, a plot of $\log p$ against $1/T$ should give a straight line provided ΔH is constant over the range of temperature investigated. ΔH may be determined from the slope of this graph.

Apparatus and Chemicals

Vapour pressure apparatus, glass bulbs, water pump, thermometer graduated in $0 \cdot 1°C$, large beaker and benzene.

Method

A small glass bulb is filled with benzene by warming it in an oven and then immersing it in a beaker of benzene. It is then attached to the thermometer so that the bulb containing the benzene is close to the bulb of the thermometer. The thermometer is fitted into the bolt head flask which should contain enough water to ensure that the entire glass bulb is submerged. The flask is immersed in a large beaker of water fitted with a stirrer.

The apparatus is warmed very gently until the temperature is about 30°C, and a few minutes are allowed for the attainment of thermal equilibrium. By adjusting the water pump and air leak in conjunction, the pressure in the system is reduced until bubbles just emerge from the jet of the bulb. The pressure indicated by the manometer is read. The pressure in the system is further adjusted until the stream of bubbles from the bulb just stops. Once again the pressure indicated by the manometer is read. The mean of the two manometer readings is subtracted from the atmospheric pressure and the result is the vapour pressure of benzene at the temperature of the experiment.

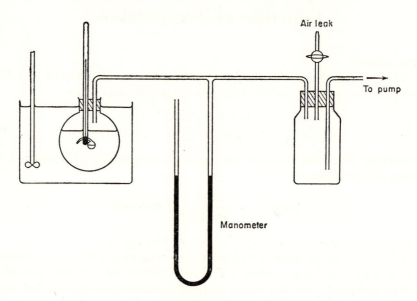

FIG. 1. Vapour pressure apparatus.

Further measurements are made at intervals of 5–7°C up to a temperature of about 75°C. As it is undesirable that the water should boil, the pressure in the system should be steadily increased as the temperature is raised. It is essential, however, that water should not be allowed to suck back into the bulb at any time during the experiment. Thus, the pressure should be continually adjusted to ensure that the water does not boil and that a steady stream of bubbles emerges from the bulb at all times other than when a measurement is being made.

A graph of $\log p$ against $1/T$ is plotted and ΔH is determined from the slope.

Heat of Solution from Solubility

Discussion

The variation of the solubility of a substance with temperature is given by the relation

$$\frac{d \ln S}{dT} = \frac{\Delta H}{RT^2} \tag{1}$$

where S is the solubility of the substance and ΔH its heat of solution. This relationship may be regarded as a form of the van't Hoff isochore.

Assuming ΔH is constant between the temperatures T_1 and T_2 and integrating equation (1) between these limits

$$\ln\frac{S_1}{S_2} = \frac{\Delta H}{R}\left[\frac{1}{T_2} - \frac{1}{T_1}\right]$$

or

$$\log\frac{S_1}{S_2} = \frac{\Delta H}{2\cdot303R}\left[\frac{T_1 - T_2}{T_1 T_2}\right] \tag{2}$$

where S_1 is the solubility at the absolute temperature T_1, and S_2 is the solubility at the absolute temperature T_2.

If the solubility of a substance is determined at two different temperatures its heat of solution may be calculated by application of equation (2). If more accurate data are required solubilities may be determined at several different temperatures and ΔH may be obtained from a plot of $\log S$ against $1/T$.

Apparatus and Chemicals

Eight 250 ml conical flasks, thermostats at 25°C, 35°C, 45°C and 55°C, 0·1 N sodium hydroxide, benzoic acid, 25 ml pipettes, cotton wool and rubber tubing.

Method

About 1 g of benzoic acid is placed in each of two 250 ml conical flasks and 100 ml of hot water at about 70°C is added to each. One flask is placed in the thermostat at 25°C and the other flask is placed in the thermostat at 35°C. In addition, about 2 g of benzoic acid is placed in each of two further conical flasks, 100 ml of hot water is added to each, one of these flasks being placed in the thermostat at 45°C and the other in the thermostat at 55°C. The four solutions are allowed to come to thermal equilibrium, the time required being about half an hour if the solutions are stirred from time to time.

A small piece of rubber tubing containing a loose wad of cotton wool is attached to a 25 ml pipette and 25 ml samples of the solutions are withdrawn. The sampling operation should be rapid. The solid should be allowed to settle before an aliquot is taken and care should be exercised to avoid clogging the filter due to excessive ingression of solid particles. To prevent the crystallization of the solutions at the higher temperatures in the pipette, the pipette may be warmed before use.

The filter is removed from the pipette and the sample is run into a conical flask. Any benzoic acid remaining in the pipette should be washed into the flask. The samples from the solutions at 25°C and 35°C should be titrated with 0·02 N sodium hydroxide solution made by accurately diluting the stock solution. The samples from the solutions at the higher temperatures may be titrated with the 0·1 N sodium hydroxide. Phenolphthalein is used as an indicator.

Assuming that the densities of the solutions are unity, the solubility of benzoic acid in grams per 100 g of water at each temperature is calculated. The heat of solution of benzoic acid may then be calculated from equation (2) or from a plot of $\log S$ against $1/T$.

Heat of Combustion by Bomb Calorimetry

Discussion

In thermochemical calculations standard heats of formation provide many useful data. The standard heats of formation of many organic compounds may be calculated from a knowledge of the heat of combustion of the compound and the standard heats of formation of carbon dioxide and water. A convenient way of determining these heats of combustion is with a bomb calorimeter, in which the substance is burnt in an atmosphere of compressed oxygen in a closed vessel. In this way the reaction is carried out at constant volume and the heat evolved is therefore equal to the decrease in the internal energy of the system. The heat of combustion at constant pressure, the enthalpy change of the reaction, may then be calculated from the relation

$$\Delta H = \Delta U + \Delta n R T$$

where Δn is the increase in the number of moles of gas during the reaction.

Apparatus and Chemicals

Bomb calorimeter outfit, stop clock, benzoic acid, naphthalene and 0·1 N sodium hydroxide solution.

Method

There are several makes of bomb calorimeter available and the method of operation will vary in detail depending on the make of calorimeter and according to whether it is of the isothermal or adiabatic type. In all cases the manufacturer's instruction manual should be consulted for the detailed operation of the apparatus, but a general description will be given here for an isothermal calorimeter in order to illustrate the principles involved.

The first objective is to determine the water equivalent of the calorimeter and accessories. This is done by burning a substance of accurately known heat of combustion such as benzoic acid.

About 1 g of benzoic acid is compressed into a pellet and weighed. The pellet is placed in the crucible of the bomb and the firing wire, either iron or platinum, is attached to the electrodes inside the bomb and in contact with the pellet. The firing wire may be incorporated in the pellet if desired. One or two millilitres of water is introduced into the body of the bomb which is then assembled. Oxygen is now admitted to the bomb until the pressure inside is about 20 atm.

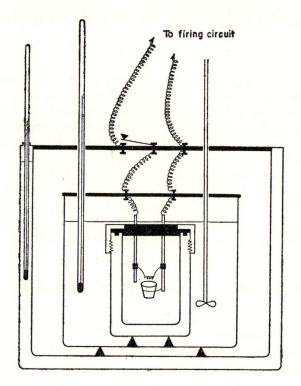

To firing circuit

FIG. 1. Isothermal bomb calorimeter.

The bomb is placed in position in the calorimeter vessel and a measured volume of water, sufficient just to submerge the upper surface of the bomb cap, is added. The weight of water used should not vary by more than 1 g in subsequent experiments. The temperature of this water should be about 2°C lower than the temperature of the water contained in the outer jacket of the apparatus. If the bomb is not gas-tight the fault will now be detected by the appearance of bubbles at any leak.

The calorimeter vessel is transferred to the outer jacket, the stirrer, thermometers and covers being placed in position and the bomb is connected to the firing circuit. The stirrer is started and 5 min should elapse before any readings are taken. After this preliminary period of 5 min has elapsed, the temperature of the water in the calorimeter vessel is read as accurately as possible at intervals of 1 min for the remainder of the experiment.

On the fifteenth minute from the start (i.e. 10 min after temperature readings have begun) the firing key is depressed. The temperature will now rise rapidly, reach a maximum and fall again slowly as indicated in Fig. 2. When the temperature is falling regularly readings should be taken for a further 10 min. The temperature of the water in the outer jacket should be checked occasionally during the experiment.

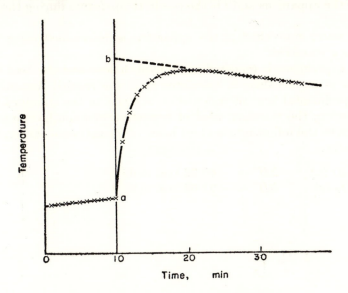

FIG. 2. Variation of temperature with time.

The bomb is then removed from the calorimeter vessel, the pressure gently released and the bomb dismantled. After ensuring that combustion has been complete the bomb is rinsed out with a little distilled water and the washings titrated with 0·1 N sodium hydroxide solution to determine the amount of nitric acid formed from the nitrogen of the air originally present in the bomb. If the firing wire was of iron any uncombusted iron wire will have to be weighed to determine the amount which has been oxidized.

The temperature rise is obtained by extrapolation of the latter part of the graph to the point *b*, the actual rise corresponding to the interval *ab*.

Given that the heats of combustion of benzoic acid and iron are 6·318 kcal g^{-1} and 1·6 kcal g^{-1}, respectively, and that the heat of formation of dilute aqueous nitric acid from nitrogen, oxygen and water is 13·8 kcal mole^{-1} the water equivalent of the calorimeter may now be calculated assuming that the specific heat capacity of water is 1 cal °C^{-1} g^{-1}.

If x g of benzoic acid and y g of iron are oxidized, the amount of nitric acid formed being z moles, then

$$6318x + 1600y + 13,800z = t(w_1 + w_2)$$

where w_1 is the weight of water in the calorimeter, w_2 the water equivalent of the apparatus and t is the rise in temperature during the experiment.

The water equivalent of the apparatus is determined using benzoic acid as a standard.

The experiment is repeated using naphthalene and the heat of combustion of naphthalene is calculated from the results. Assuming that the experimental heat of combustion is equal to the standard heat of combustion, the standard heat of formation of naphthalene is calculated given the following standard heats of formation of carbon dioxide and water:

H_2O (l.) $\Delta H° = -68·32$ kcal mole^{-1}
CO_2 (g) $\Delta H° = -94·03$ kcal mole^{-1}

An Equilibrium Constant by the Distribution Method

Discussion

The reversible reaction

$$KI + I_2 \rightleftharpoons KI_3$$

occurs in aqueous solution and according to the law of mass action the equilibrium constant K_c is given by:

$$K_c = \frac{[KI_3]}{[KI][I_2]}$$

The equilibrium may be investigated by studying the distribution of iodine between an organic solvent and water, followed by a similar study for the distribution of iodine between the same organic solvent and an aqueous solution of potassium iodide. The first study enables the distribution coefficient k to be calculated from the relationship

$$\frac{C_{org}}{C_{aq}} = k$$

where C is the concentration of the iodine (determined by titration).

Since the distribution law only applies to the species common to both layers, the concentration of free iodine in the aqueous potassium iodide layer, $C_{KI,aq}$, can be determined from the relationship

$$C_{KI,aq} = \frac{C_{org}}{k}$$

The iodine combined with potassium iodide to form KI_3 may now be determined since the total iodine in the aqueous potassium iodide layer may be obtained by titration.

The amount of iodide which has combined with the iodine can then be found and as the original concentration of iodide is known the amount of uncombined iodide may be obtained by difference. Hence K_c may be calculated.

Apparatus and Chemicals

Three 250 ml glass stoppered bottles, thermostat, 0·1 N potassium iodide, iodine, carbon tetrachloride and 0·1 N sodium thiosulphate.

Method

Twenty millilitres of a saturated solution of iodine in carbon tetrachloride is shaken with 200 ml of water in one of the glass stoppered bottles. The bottle is immersed in the thermostat at 25°C and the bottle and contents allowed to attain thermal equilibrium. The same procedure is followed using 15 ml and 10 ml portions of the saturated solution of iodine diluting the former with 5 ml of carbon tetrachloride and the latter with 10 ml of carbon tetrachloride.

A 50 ml portion of the aqueous layer and a 5 ml portion of the organic layer of each mixture is analysed using 0·1 N or 0·01 N sodium thiosulphate depending upon the amount of iodine present in the samples. A small quantity of concentrated potassium iodide solution is added to the organic solution to ensure complete extraction of the iodine.

The whole procedure is repeated with a 0·1 N solution of potassium iodide in place of water and using 20 ml, 15 ml and 10 ml portions of a saturated solution of iodine in carbon tetrachloride diluted as before.

Distillation of an Azeotropic Mixture with a Minimum Boiling Point

Discussion

Binary liquid mixtures showing a large positive deviation from Raoult's law exhibit a maximum in the vapour pressure vs. composition curve, Fig. 1.

For this reason, the mixture, the composition of which corresponds to the maximum vapour pressure, will have the minimum boiling point of any mixture of the two components. Furthermore, it will be seen from Fig. 2 that the liquidus and vapourus curves coincide at the temperature of the minimum boiling point.

The number of degrees of freedom F, of a system consisting of C components and P phases, is given by:

$$F = C - P + 2$$

For a binary system at constant pressure this reduces to

$$F = 3 - P$$

On application of the phase rule to the point M an identity restriction must be introduced by subtracting one degree of freedom since the compositions of the liquid and the vapour are the same at this point. Thus, at constant pressure, M is an invariant point, i.e. $F = 0$.

A mixture which distils without change of composition is said to be azeotropic. The mixture of composition corresponding to the point M is thus an azeotrope.

Apparatus and Chemicals

Abbé refractometer, thermometer 70–110°C graduated in 0·1°C, stoppered weighing bottles, 500 ml QQ three-necked flask, B19 stopper, B14 thermometer pocket, modified QQ C 1/12 condenser, two semi-micro pipettes, heating mantle, isopropanol and benzene.

Method

A series of mixtures of known composition (mole per cent of isopropyl alcohol in benzene) are prepared and their refractive indices measured. The refractive indices of pure isopropyl alcohol and pure benzene are also measured. A plot of refractive index against mole fraction of isopropyl alcohol in an isopropyl–benzene mixture is prepared (Graph 1). This data can be conveniently determined and correlated while the distillation is being carried out. Alternatively, time will be saved if a previously prepared graph is obtainable in the laboratory.

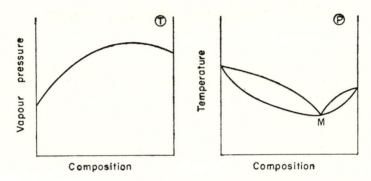

FIG. 1. Vapour pressure vs. composition.

FIG. 2. Temperature vs. composition.

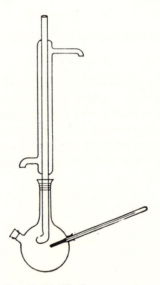

FIG. 3 Distillation apparatus.

A quantity of pure benzene (125 ml) is placed in the flask and its boiling point determined. The thermometer is read with the aid of a lens. A smaller quantity (5 ml) of isopropyl alcohol is added to the benzene, the mixture is refluxed for several minutes and the temperature is recorded. The flask is then raised from the mantle and when boiling has ceased a pipette is used to take up a few drops of the distillate from the small cup at the lower end of the condenser. A second pipette is used to take a sample of the residue in the flask. The refractive indices and hence the compositions of each are determined. This procedure is repeated after adding successively 5, 5, 5, 15, 25 and 25 ml of isopropyl alcohol to the benzene in the flask.

The flask is emptied, rinsed with isopropyl alcohol and 50 ml of isopropyl alcohol is added to it. The boiling point of the isopropyl alcohol is determined. Samples of the residues and distillates are taken, as before, after adding successively, five 10 ml samples of benzene to the alcohol.

Two further graphs are prepared. A plot of boiling point against mole fraction of isopropyl alcohol in the residue (Graph 2a), and a plot of boiling point against mole fraction of isopropyl alcohol in the distillate (Graph 2b). The minimum boiling point and the composition of the azeotropic mixture are noted and recorded.

At the conclusion of the experiment all benzene–isopropyl alcohol mixtures should be emptied into a residue bottle. Caution: these liquids are inflammable.

Molecular Weight of a Liquid by Steam Distillation

Discussion

If two immiscible liquids A and B are distilled from the same vessel, since one does not affect the vapour pressure of the other, they will pass over in the ratio of their vapour pressures until one of them is completely distilled. Furthermore, the vapour pressure of the mixture will be equal to the sum of the vapour pressures of the individual components, i.e.

$$p_{total} = p_A + p_B$$

At the boiling point $(p_A + p_B)$ will then be equal to the atmospheric pressure. At this temperature the mixture distils in the molecular proportions p_A/p_B. Hence if w_A and w_B are the weights in grams of A and B in the distillate, then

$$\frac{p_A}{p_B} = \frac{w_A/M_A}{w_B/M_B}$$

i.e.
$$M_A = \frac{p_B}{p_A}\frac{w_A}{w_B}M_B$$

where M_A and M_B are the molecular weights of A and B respectively. Hence the molecular weight of chlorobenzene (A) can be found by a method of steam distillation.

Apparatus and Chemicals

Steam distillation assembly, thermometer 70°–120°C graduated to 0·1°C, 2 × 100 ml graduated cylinders, separating funnel and chlorobenzene.

Method

The thermometer is calibrated in the steam distillation apparatus; the true b.p. for water at the barometric pressure of the laboratory can be obtained from tables or calculated by assuming a change of 1°C for a pressure change of 27 mm. The flask is emptied.

44

Chlorobenzene (250 ml) and 50 ml of water are added to the flask and the steam distillation allowed to commence. Steam distillation should proceed at the rate of about 1 drop per second, and the temperature recorded every minute. The distillate is collected in four fractions and the receiver changed when about 50 ml have been collected.

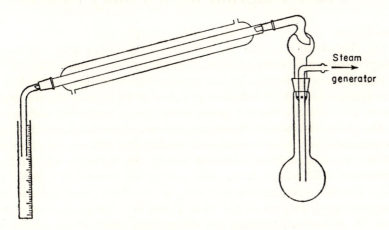

Fig. 1. Steam distillation apparatus.

Distillation is stopped when the temperature commences to rise abruptly to the b.p. of water. Each fraction is collected in a measuring cylinder.

The results are recorded in tabular form under the headings:

Observed temperature of distillation recorded every minute (°C)	Total distillate (ml)	Volume of chlorobenzene (ml)	$\dfrac{v_A}{v_B}$	$\dfrac{w_A}{w_B} = \left(\dfrac{v_A}{v_B}\right)1{\cdot}106$

The specific gravity of chlorobenzene at 15°C = 1·106.

Phase Diagram for a Binary System

Discussion

On the phase diagram for a two-component system the range of temperatures and compositions in which there exists solid and liquid in equilibrium is defined by the solidus and liquidus curves. If a solid of a given composition is heated it will start to melt at the temperature where the ordinate for the given composition meets the solidus curve on the phase diagram. In the region between the solidus and the liquidus curves, solid will be in equilibrium with liquid but at the temperature where the composition ordinate meets the liquidus curve the last particle of solid will melt and the system will consist of a liquid phase only. The first temperature is known as the thaw point and the second temperature is known as the melt point for that particular composition. If these temperatures are determined for various compositions the phase diagram for the system may be constructed. This is the thaw–melt method of thermal analysis.

Apparatus and Chemicals

Ignition tubes, melting point tubes, pestle and mortar, melting point apparatus, p-nitrotoluene and naphthalene.

Method

A mixture of naphthalene and p-nitrotoluene is made up such that the mole fraction of the naphthalene in the mixture is about 0·1. The mixture is melted with a bunsen burner to ensure thorough mixing and the melt is then allowed to solidify. The solid mixture is placed in a mortar and ground thoroughly to make it as homogeneous as possible. A small amount of the finely ground mixture is placed in a melting point tube and gently warmed in an apparatus for the determination of melting points. The temperature at which the first drop of liquid appears is noted as also is the temperature at which the system becomes completely molten.

The experiment is repeated for mixtures having a mole fraction of naphthalene of 0·2, 0·3, 0·4, 0·5, 0·6, 0·7, 0·8 and 0·9. The melting points of pure naphthalene and pure p-nitrotoluene are also determined.

From the results the phase diagram for the system may be constructed.

The Variation of Miscibility with Temperature

Discussion

For certain pairs of liquids there are temperature limits to their miscibility. For such systems at constant pressure the phase rule reduces to

$$F = 3 - P$$

Thus whilst two layers exist in equilibrium P equals 2 and the system is completely defined either by temperature or composition.

A graph may therefore be plotted indicating the temperature limits of miscibility for a series of mixtures of known composition.

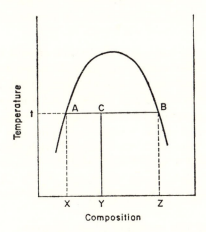

FIG. 1. Partial miscibility of two liquids.

The tie-line principle may then be applied to this graph to obtain the composition and relative proportions of each phase of a particular sample at a given temperature. Thus, with reference to Fig. 1, at temperature t, a sample having composition Y exists as two layers A and B. A and B have the composition X and Z, respectively, and the relative proportions by weight are

$$\frac{A}{B} = \frac{BC}{AC}$$

Apparatus and Chemicals

Nine boiling tubes, burette, 400 ml beaker, glycerol and *m*-toluidine.

Method

A series of nine synthetic samples of glycerol and *m*-toluidine are prepared using the amounts shown in Table 1. The mixtures are made up by weighing the glycerol into a tube and adding the *m*-toluidine from a burette.

TABLE 1

Tube	Glycerol (g)	*m*-Toluidine (g)
1	4	25
2	5	25
3	6	25
4	10	35
5	10	20
6	20	10
7	6·6	2
8	6	1·5
9	9·9	2·2

With the exceptions of tubes 5 and 6 which should be cooled in iced water, the tubes are cooled under running cold water and stirred gently with a thermometer until the contents become clear. Tubes 1, 2, 3, 4, 7, 8 and 9 are warmed in water and tubes 5 and 6 are allowed to warm up in water to which some ice has been added. The solutions are again stirred and the temperature at which the liquid just becomes turbid is noted (t_1°C). Each tube is further heated with shaking in a water bath or, for tubes 5 and 6, an oil bath and the temperatures at which the mixtures become clear again are noted (t_2°C). The temperatures t_1 and t_2 are plotted against the weight percentage compositions of the mixtures.

If there are gaps in the resulting phase diagram, further mixtures should be made up of such compositions as to enable the diagram to be completed.

A Study of the Ternary System:
Benzene–Acetic Acid–Water

Discussion

For a three-component system at constant temperature and pressure the phase rule reduces to

$$F = 3 - P$$

The composition of such a system may be expressed in terms of the co-ordinates of an equilateral triangle, each apex corresponding to a pure component.

Thus the point X in Fig. 1 corresponds to a composition of 25 per cent A, 50 per cent B and 25 per cent C.

The components benzene and water are essentially immiscible hence two layers will be obtained on mixing. The third component, viz. acetic acid, is soluble in each layer and on addition to a mixture of benzene and water distributes itself between these layers. The composition of the layers changes as more acetic acid is added. Under these conditions, P equals 2, and the system is completely defined by the composition. Thus if the points corresponding to these compositions are plotted, a smooth curve may be drawn indicating the limits of miscibility of water–benzene mixtures in acetic acid.

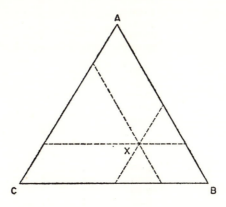

FIG. 1. Representation of composition on a triangular diagram.

It will be noted that as more and more acetic acid is added, the water and benzene solutions become more and more miscible until complete miscibility is achieved.

This solubility curve is known as a binodal curve. Any mixture within the area enclosed by this binodal curve and the base of the triangle will resolve itself into two liquid layers and any mixture outside the area will form only one liquid layer. The position of this curve changes with temperature.

Apparatus and Chemicals
One millilitre pipette, 5 ml pipette, benzene and acetic acid.

Method
Five millilitres of benzene is placed in a 100 ml flask. One millilitre of water is added from a burette and glacial acetic acid is slowly added from a second burette until a clear solution is just obtained on vigorously shaking the mixture. The volume of acetic acid which has been added

TABLE 1

Series	Water (ml)	(moles)	(mole fraction of water)	Benzene (ml)	(moles)	(mole fraction of benzene)	Acetic acid (ml)	(moles)	(mole fraction of acetic acid)	Total moles
A	1·00			5·00						
	1·00			10·00						
	1·00			15·00						
	1·00			20·00						
	1·00			30·00						
	1·00			35·00						
B	3·00			5·00						
	6·00			5·00						
	10·00			5·00						
	15·00			5·00						
	21·00			5·00						
C	25·00			2·00						
	50·00			2·00						
	75·00			2·00						
	100·00			2·00						
	125·00			2·00						
	150·00			2·00						

is then recorded. Successive volumes of 5·0 ml of benzene are added to the mixture. After each addition of benzene, acetic acid is run in and the mixture shaken vigorously until the system becomes homogeneous. The results (Series A) are recorded as indicated in Table 1.

In Series B, 5·00 ml of benzene is placed in a 100 ml conical flask and treated successively with 3, 3, 4, 5 and 6 ml of water and titrated with acetic acid after each addition. In the Series C experiments 2·00 ml of benzene is placed in a dry 250 ml conical flask and is treated successively with six 25·00 ml portions of water and titrated with acetic acid after each addition.

The molar volumes of the components are determined from the densities of water, benzene and acetic acid which at room temperature are 1·000, 0·872, 1·049 g ml⁻¹, respectively. The composition of each mixture which just reaches the point of complete miscibility can then be calculated in terms of mole fractions and recorded in Table 1.

In order to determine one of the tie lines the following procedure may be adopted. The two layers of a mixture of 25 ml benzene, 7 ml water and 17 ml acetic acid are separated with a separating funnel and each layer is weighed. The overall mole fraction composition of the mixture Z, and the ratio of the weight of the water rich layer to the weight of the benzene rich layer w_1/w_2, are calculated.

After the main curve has been drawn on the diagram, a straight line is drawn through the point Z, to intersect the benzene rich and water rich sides of the curve at P and Q respectively, such that $PZ/ZQ = w_2/w_1$. The tie line is drawn between P and Q.

Note. If time is limited, alternate titrations in series B and C of Table 1 may be carried out and the remainder omitted.

C

The Thermal Analysis of a Hydrate by the Differential Couple Method

Discussion

Copper sulphate pentahydrate dehydrates according to the following reactions:

$$CuSO_4 . 5H_2O \rightleftharpoons CuSO_4 . 4H_2O + H_2O$$

$$CuSO_4 . 4H_2O \rightleftharpoons CuSO_4 . 3H_2O + H_2O$$

$$CuSO_4 . 3H_2O \rightleftharpoons CuSO_4 . H_2O + 2H_2O$$

$$CuSO_4 . H_2O \rightleftharpoons CuSO_4 + H_2O$$

The phase rule for such a system at constant pressure reduces to

$$F = 3 - P$$

The dehydration may be studied by using a differential thermo-couple[1] with one junction placed in hydrated copper sulphate and the other junction in sodium chloride. If this couple is connected to a sensitive galvanometer and the two salts heated in the same enclosure then the galvanometer will indicate no passage of current as long as no phase reaction takes place in the hydrated copper sulphate.

However, at the temperature at which one of the above phase reactions takes place, three phases are present, i.e. $P = 3$, hence $F = 0$. There will therefore be a temperature arrest until the phase reaction is complete. Since sodium chloride suffers no phase transitions there will be no temperature arrest in the sodium chloride. The galvanometer will thus indicate a passage of current at the temperature of each phase reaction of the hydrated copper sulphate.

It will be noted that the dehydration of the monohydrate involves no temperature arrest since the single molecule of water is attached to the sulphate radical by physical forces and is thus lost gradually as the temperature is raised.

Apparatus and Chemicals

Dewar flask, copper block, heater and variable resistance, thermo-couple, two thin-walled glass tubes, stirrer, sensitive galvanometer, accurate thermometer, copper sulphate pentahydrate and sodium chloride.

Method

A copper block is drilled with two holes into which fit two thin-walled glass tubes. To one of the tubes is added approximately 1·5 g powdered copper sulphate pentahydrate and to the other a similar

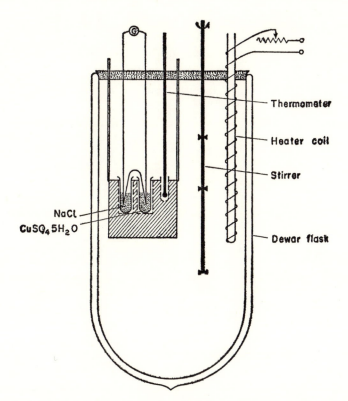

FIG. 1. Heating unit for thermal analysis by differential couple method.

quantity of powdered sodium chloride. The differential thermocouple is embedded in the two salts and each end connected through a sensitive galvanometer. A third hole in the block contains the bulb of an accurate thermometer. The copper block and contents are suspended in a Dewar flask together with a stirrer and heater coil (Fig. 1).

By adjusting the supply to the heater, by means of the variable resistance included in the circuit, the temperature inside the Dewar flask is raised rapidly to 90°C and subsequently at the rate of 0·5°C to 1°C per min. The temperatures are noted when the galvanometer shows a deflection.

Greater accuracy can be achieved if a calibrated thermocouple is used to measure the temperature instead of a thermometer.

[1] TAYLOR and KLUG, *J. Chem. Phys.*, **4**, 601 (1936).

A Transition Temperature by a Solubility Method

Discussion

At atmospheric pressure the eutectic system of sodium bromide and water forms a compound having an incongruent melting point. The definition of such a compound is that it is incapable of existence in the presence of a solution of the same composition.

The phase diagram for this type of system is shown in Fig. 1. In this particular case where one of the components A is water, the curves CD and DB may be regarded as the solubility curves for sodium bromide dihydrate and anhydrous sodium bromide, respectively. The transition point D can therefore be obtained as the point of intersection of the two curves.[1]

The transition point is invariant at constant pressure since under these conditions the phase rule reduces to

$$F = 3 - P$$

and at D the two solid phases coexist with saturated solution, i.e.

$$P = 3$$

Apparatus and Chemicals

Five-millilitre container pipette, sodium bromide dihydrate ($NaBr . 2H_2O$) and 0.1 M silver nitrate.

Method

A saturated solution of sodium bromide at 75°C is prepared by placing approximately 250 g of sodium bromide dihydrate and 100 ml water in a small flask. The flask is placed in a water bath and 5 ml samples withdrawn every 5°C–10°C over the range 25°C–75°C. The first solution to be analysed is at 75°C. The solution is then allowed to cool to the lower temperatures thus ensuring saturation at each temperature. The samples are removed by a pipette fitted with a cotton wool-packed guard and run into a 100 ml graduated flask. Any crystals deposited on the inside of the pipette are carefully rinsed into the 100 ml

55

graduated flask before making the solution up to the mark. The diluted sodium bromide solution is titrated with 0·1 M silver nitrate solution.

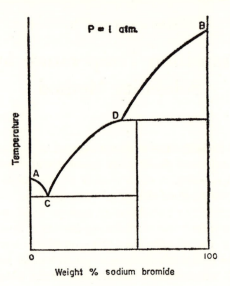

FIG. 1. Temperature vs. composition phase diagram for sodium bromide and water at atmospheric pressure.

A graph of the volume of 0·1 M silver nitrate (ml) vs. the temperature is plotted and the solubility in grams per 100 ml of sodium bromide dihydrate (NaBr . $2H_2O$) at the transition point calculated.

[1] R. T. Bottle and G. N. Copley, *School Sci. Rev.*, **41**, 423 (1960).

Verification of the Lambert–Beer Law

Discussion

When light falls on a homogeneous medium, it is partly reflected, partly absorbed, and partly transmitted. The relationship between the intensity of the incident light and that of the transmitted light was investigated by Lambert and extended to solutions by Beer. According to Lambert the rate of decrease of intensity of the light with thickness of absorbing medium, $-dI/dl$, I being the intensity at thickness l (cm), is proportional to the intensity of the light at point l, i.e.

$$-dI/dl = k'I \qquad (1)$$

The constant k' is called the absorption coefficient and it is characteristic of the absorbing medium.

On rearranging equation (1) and integrating between the limits $l = 0$ corresponding to I_0, and $l = 1$ corresponding to I, we have

$$\int_{I_0}^{I} dI/I = - \int_{0}^{l} k' \, dl$$

$$\ln I/I_0 = -k'l \qquad (2)$$

$$I/I_0 = e^{-k'l}$$

$$I/I_0 = 10^{-kl}$$

where $k = k'/2 \cdot 203$. The constant k is known as the extinction coefficient.

The fraction I/I_0 is the fraction of light transmitted and is called the transmittance T. The common logarithm of the reciprocal of this ratio, i.e. $\log(I_0/I)$, is known as the absorbance A. Note the relationships

$$A = \log(I_0/I) = -\log T = kl \qquad (3)$$

For an absorbing substance dissolved in a transparent solvent, the decrease in intensity is also proportional to the concentration of the solution, hence

$$-dI/dl = \epsilon'Ic$$

where c is the concentration in moles per litre and ϵ' is the molar absorption coefficient. This constant is characteristic of the solute. For solutions, the transmission is related to the molar absorption coefficient by

$$-\ln T = \epsilon' c l$$

The corresponding relationship involving common logarithms is

$$A = -\log T = \epsilon c l \qquad (4)$$

where ϵ is called the molar absorptivity.

Apparatus and Chemicals

Spectrophotometer, e.g. Unicam S.P.600, test tubes, test tube rack, 5×100 ml graduated flasks, 0·880 ammonia solution and copper sulphate.

Method

A 0·1 M solution of copper sulphate is prepared. Solutions of various concentrations of the blue cuprammonium ion are prepared by mixing 5 ml of 0·880 ammonia, previously diluted to 10 ml with water, with 2, 4, 6, 8, 10, 12, 14, 16, 18, 20, 22 and 24 ml of the copper sulphate solution. The final volume is adjusted to 100 ml by addition of a further volume of water.

At a wavelength of 600 mμ the percentage transmission and the absorbances of these solutions are measured. The manufacturer's manual should be consulted for the operating instructions. Graphs of (a) percentage transmission vs. concentration, and (b) absorbance vs. concentration can be drawn from the data. As indicated by equation (4), if the Lambert–Beer relationships are valid for this solution, graph (a) should be exponential, whereas graph (b) should be linear.

Composition of a Complex Ion in Solution

Discussion

Spectrophotometry provides a convenient method for the investigation of the ionic equilibrium between simple and complex ions.[1] Consider the formation of a complex ion in solution from a metallic ion A and a second particle B (anion or molecule). The equilibrium may be written

$$A + nB = AB_n$$

If the concentrations of A and B are each M moles/l., then on mixing x l. of B with $(1-x)$ l. of A (where x is less than unity), the concentration of the ionic species in the solution will be given by

$$c_A = M(1-x) - c_{AB_n} \qquad (1)$$

$$c_B = Mx - nc_{AB_n} \qquad (2)$$

These concentrations are also related to each other by the equation

$$c_A \cdot c_B^n = Kc_{AB_n} \qquad (3)$$

where K is the instability constant. For a plot of C_{AB_n} vs. x, the condition for a maximum is

$$dc_{AB_n}/dx = 0 \qquad (4)$$

Differentiating equations (1), (2) and (3) with respect to x and combining the resulting equations with equations (1), (2), (3) and (4) gives

$$n = x/(1-x) \qquad (5)$$

The molar absorptivity ϵ is related to the absorbance A by the relationship

$$A = \epsilon c l$$

where c is the concentration (g moles/l.) and l the length of the path of light through the solution. Hence for the mixture

$$A = l(\epsilon_1 c_A + \epsilon_2 c_B + \epsilon_3 c_{AB_n}) \qquad (6)$$

where ϵ_1, ϵ_2 and ϵ_3 are the molar absorptivities of the species, A, B and AB_n. If no reaction had occurred on mixing, the absorbance A' would have been

$$A' = l[\epsilon_1 M(1-x) + \epsilon_2 Mx] \tag{7}$$

Hence

$$A - A' = t(\epsilon_1 \cdot c_A + \epsilon_2 \cdot c_B + \epsilon_3 c_{AB_n} - \epsilon_1 M(1-x) - \epsilon_2 Mx) \tag{8}$$

The condition for maximum or minimum is that $\mathrm{d}(A-A')/\mathrm{d}x = 0$, and it can be shown that when $\epsilon_3 > \epsilon_1$ there is a maximum for this excess absorbance $(A-A')$ and here c_{AB_n} is also a maximum. If a plot of $(A-A')$ vs. x is drawn, the value of x for the maximum value of $(A-A')$ can be determined. This value of x enables n to be calculated from equation (5).

A simple example of the use of this method is to determine the composition of chromate ions in acid solution.

$$H^+ + n\,CrO_4^{2-} \leftrightharpoons H(CrO_4)_n{}^{(1-2n)-}$$

(The formation of dichromate ions due to the reaction

$$2HCrO_4{}^- = Cr_2O_7{}^- + H_2O$$

does not interfere with the method. The maximum for the dichromate ion coincides with the maximum for the parent ion.)

Apparatus and Chemicals

Spectrophotometer, e.g. Unicam SP600, two burettes, 0·1 M solution of potassium chromate, 0·1 M hydrochloric acid.

Method

A series of mixtures of the acid and chromate solutions are prepared by adding x ml of the 0·1 M hydrochloric acid to $(10-x)$ ml of 0·1 M potassium chromate solution. The absorbance of this mixture is determined at 525 mμ. The absorbance of the pure chromate solution is determined at the same wavelength.

The excess absorbance is calculated.

A plot of $(A-A')$ vs. x is prepared and the value of x corresponding to the maximum on the graph is noted. The value of n is calculated from equation (5).

[1] W. C. Vosburgh and G. R. Cooper, *J. Amer. Chem. Soc.*, **63**, 437 (1941).

A Velocity Constant by a Titration Method

Discussion

The rate of a chemical reaction is proportional to the concentrations of the reactants and is usually expressed in terms of the concentration of one of the reactants or products of the reaction, i.e.

$$-\frac{dc}{dt} \quad \text{or} \quad \frac{dx}{dt}$$

where c is the concentration of one of the reactants and x is the concentration of one of the products at a time t.

For the general reaction

$$A + B + C + \ldots \rightarrow \text{products}$$

the rate equation may be expressed in the form

$$-\frac{dc}{dt} = k(A)^{n_1}(B)^{n_2}(C)^{n_3} \ldots \tag{1}$$

where k is the velocity constant for the reaction. The order of the reaction n is the sum of the exponents in the rate equation, i.e.

$$n = n_1 + n_2 + n_3 + \ldots$$

For the second-order reaction involving the oxidation of iodide with persulphate

$$2I^- + S_2O_8^{2-} \rightarrow I_2 + 2SO_4^{2-}$$

the rate equation may be written

$$\frac{dx}{dt} = k_2(a-x)(b-2x) \tag{2}$$

where a and b are the initial concentrations of persulphate and iodide respectively.

If one of the reactants is present in excess such that its concentration remains sensibly constant during the course of the reaction, the reaction will follow first-order kinetics.

61

Suppose the concentration of iodide in the above reaction is so large that it will be approximately unaltered during the reaction. Equation (2) will reduce to

$$\frac{dx}{dt} = k_2(a-x)b \tag{3}$$

Integrating and applying the condition that when $t = 0$ and $x = 0$, we have

$$bk_2 = \frac{1}{t}\ln\frac{a}{a-x}$$

or

$$\ln(a-x) = \ln a - k't \tag{4}$$

where $k' = b \cdot k_2$.

Hence a plot of $\log(a-x)$ vs. t will be a straight line and k' is determined from the slope.

Apparatus and Chemicals

Thermostat, stop watch, three conical flasks with ground glass stoppers, saturated potassium persulphate solution, 0.4 M potassium iodide solution and 0.01 N sodium thiosulphate solution.

Method

Fifty millilitres of the 0.4 N solution of potassium iodide is delivered into a weighted conical flask from a pipette and the flask suspended in a thermostat at 25°C. Twenty millilitres of the saturated solution of potassium persulphate is diluted with 80.0 ml of water and 50.0 ml of the resulting solution placed in a conical flask and suspended in the thermostat. When the temperatures of the solutions have become steady at 25°C the potassium iodide solution is poured into the potassium persulphate solution and the stop watch started simultaneously. It is important to place the stopper in the flask containing the reactants as soon as they have been mixed in order to prevent the loss of iodine by volatilization from the reaction mixture. At measured and recorded intervals of time $(3, 8, 15, 20, 30, 40, 50, 60 \text{ min})$, 10.0 ml of the reaction mixture is withdrawn with a pipette and allowed to run into a large volume of water. This dilution has the effect of slowing down the reaction considerably whilst the titration is completed. Each 10.0 ml sample is titrated with freshly prepared 0.01 N sodium thiosulphate (x ml) using starch indicator.

Fifty millilitres of the potassium iodide solution is mixed with the remainder of the diluted potassium persulphate solution and the stopper is placed in the flask which is maintained at 60°C for the duration of the experiment. This solution is cooled and maintained at 25°C for at least 15 min and 10·0 ml titrated with 0·01 N sodium thiosulphate (a ml).

The results are recorded in tabular form under the headings:

Time t min	Titration a ml	Titration x ml	$(a-x)$	$\log(a-x)$

Variation of a Velocity Constant with Catalyst Concentration by a Polarimetric Method

Discussion

Glucose exists in two different optically active forms, viz. α-glucose and β-glucose. The former has a specific rotation $[\alpha]_D = +110°$ whilst the latter has a specific rotation $[\alpha]_D = +19°$. On standing, the value of the specific rotation of an aqueous solution of either of these forms changes and finally achieves an equilibrium value of $[\alpha]_D = +52·5°$ due to the presence of an equilibrium mixture of the α and β forms. This change in optical rotatory power is known as mutarotation.

Mutarotation is catalysed by both acid and alkali. The change is first order under both conditions of catalysis and as the reaction is a reversible one, the measured rate constant k is the sum of the individual forward and backward rate constants, k_1 and k_{-1} respectively.

$$\alpha \underset{k_{-1}}{\overset{k_1}{\rightleftharpoons}} \beta$$

If it is assumed that the initial concentration of the α form is a moles per litre and that after a time t the concentration of the β form is x moles per litre, then

$$dx/dt = k_1(a-x) \tag{1}$$

and

$$-dx/dt = k_{-1}x \tag{2}$$

The observed rate of the reaction is given by

$$dx/dt = k_1(a-x) - k_{-1}x \tag{3}$$

If x_e is the value of x at equilibrium, then

$$k_1(a-x_e) = k_{-1}x_e \tag{4}$$

hence

$$a = x_e\left(\frac{k_1+k_{-1}}{k_1}\right)$$

Substituting in equation (3) gives

$$\frac{dx}{dt} = k_1\left[x_e\left(\frac{k_1+k_{-1}}{k_1}\right)-x\right]-k_{-1}x$$

or

$$\frac{dx}{dt} = (k_1+k_{-1})(x_e-x)$$

Putting $(k_1+k_{-1}) = k$ and integrating

$$\int\frac{dx}{(x_e-x)} = \int k\,dt$$

$$-\ln(x_e-x) = kt+C \tag{5}$$

Applying the condition that when $t = 0$, $x = 0$ gives $C = -\ln x_e$, and substituting in equation (5) gives

$$-\ln(x_e-x) = kt-\ln x_e$$

Hence

$$-kt = \ln\frac{x_e-x}{x_e}$$

or

$$kt = 2\cdot303\log\frac{x_e}{x_e-x} \tag{6}$$

If α_0 is the initial angle of rotation, α_∞ the final angle of rotation and α_t the angle of rotation at a time t, then x_e is proportional to $(\alpha_0-\alpha_\infty)$ and (x_e-x) is proportional to $(\alpha_t-\alpha_\infty)$. Substitution of these values in equation (6) gives

$$kt = 2\cdot303\log\frac{\alpha_0-\alpha_\infty}{\alpha_t-\alpha_\infty}$$

or

$$\log(\alpha_t-\alpha_\infty) = \log(\alpha_0-\alpha_\infty)-\frac{kt}{2\cdot303} \tag{7}$$

A graph of $\log(\alpha_t-\alpha_\infty)$ against t is therefore linear with a slope of $-k/2\cdot303$.

In acid solution the overall velocity constant is given by

$$k = k_{H^+}.c_{H^+}+k_{OH^-}.c_{OH^-}+k_{H_2O}$$

where k_{H^+} is the catalytic constant for the hydrogen ion, k_{OH^-} the catalytic constant for the hydroxyl ion, c_{H^+} the concentration of the hydrogen ions, c_{OH^-} the concentration of the hydroxyl ions and k_{H_2O} is the catalytic constant for water. In the case of acid catalysis the term for the hydroxyl ions is negligible and a plot of the velocity constant k against c_{H^+} is linear with a slope of k_{H^+} and an intercept of k_{H_2O}.

Apparatus and Chemicals

Bellingham and Stanley B-S model polarimeter, sodium lamp, thermostat at 25°C, circulatory pump, water-jacketed polarimeter tube, glucose, 0·2, 0·1 and 0·05 N hydrochloric acid.

Method

Ninety millilitres of water and 20 ml of the 0·2 N acid solution are placed separately in two conical flasks. The flasks are placed in the thermostat and when thermal equilibrium has been attained, 10 g of glucose are dissolved in the water and 10 ml of the acid is added. The solution is shaken and a portion is transferred to the polarimeter tube. The angle of rotation is measured as rapidly as possible and the measurement is repeated at intervals of 1 min over a period of 30 min. A portion of the solution should be kept in the thermostat for 24 hr and a final reading made. The experiment is repeated with 0·1 N and 0·05 N hydrochloric acid.

A plot of $\log(\alpha_t - \alpha_\infty)$ is made against t and the velocity constant k is calculated. A plot of k values against c_{H^+} is also drawn and k_{H^+} and k_{H_2O} are obtained.

A Velocity Constant by a Gas Evolution Method

Discussion

Benzene diazonium chloride undergoes decomposition when warmed with water according to the reaction

$$C_6H_5N_2Cl + H_2O \rightarrow C_6H_5OH + HCl + N_2$$

The course of this reaction may be followed by measuring the volume of nitrogen evolved.[1]

In the presence of excess water the reaction follows first-order kinetics and if the initial concentration of benzene diazonium chloride is a and if after a time t the concentration is $(a-x)$ then we may write

$$\frac{dx}{dt} = k(a-x)$$

or

$$\frac{dx}{(a-x)} = k \, dt \tag{1}$$

On integrating equation (1) and applying the condition that when $t = 0$, $x = 0$ we have

$$\ln(a-x) = -kt + \ln a \tag{2}$$

If v_∞ is the total volume of nitrogen evolved during the whole reaction and if v_t is the volume evolved up to the time t, then a is proportional to v_∞ and x is proportional to v_t. Substituting in equation (2) we may write

$$\ln(v_\infty - v_t) = -kt + \ln v_\infty$$

or

$$\log(v_\infty - v_t) = -\frac{kt}{2 \cdot 303} + \log v_\infty$$

A plot of $\log(v_\infty - v_t)$ against t should give a straight line of slope $-k/2 \cdot 303$.

Apparatus and Chemicals

Hirsch filter tube, gas burette, mercury seal stirrer, stop clock, aniline, concentrated hydrochloric acid and sodium nitrite.

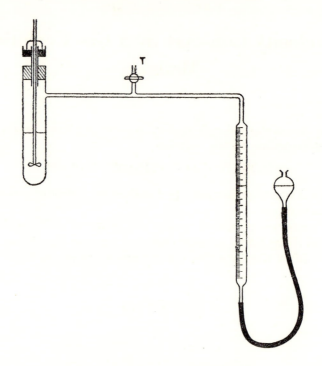

FIG. 1. Reaction vessel and gas burette.

Method

A solution of benzene diazonium chloride must first be prepared as follows. Seven grams of aniline is dissolved in 22 ml of concentrated hydrochloric acid and the resulting solution is cooled in ice water. From a dropping funnel a cold solution of 5 g of sodium nitrite in 75 ml of water is gradually added. The solution is finally made up to a litre.

About 30 ml of this solution is placed in the reaction vessel, the stirrer is fitted and the vessel is clamped in a thermostat operating at about 35°C. The system should be left open to the atmosphere at this stage by leaving the tap T, shown in Fig. 1, open. The stirrer is set in motion and 5 to 7 min are allowed to elapse so that the solution may assume the temperature of the thermostat. When thermal equilibrium has been established the tap is closed and the time is noted. This time should be taken as the time for the start of the reaction. The volume of gas evolved is read at intervals of about 15 min, the reservoir of the gas burette being continually adjusted so that the pressure in the apparatus does not depart very far from atmospheric pressure. The temperature of the gas and the barometric pressure should be noted and any marked changes during the course of the experiment should be corrected.

v_∞ is determined by immersing the reaction vessel in a beaker of hot water, cooling until the temperature of the experiment is reached and then noting the volume of gas evolved. This operation should be repeated until no further increase in volume is observed. The final volume corresponds to v_∞.

A graph of $\log(v_\infty - v_t)$ against t is plotted and the value of the velocity constant for the reaction at the temperature of the experiment is determined.

[1] CAIN and NICOLL, *J. Chem. Soc.*, **81**, 1412 (1902); **83**, 206, 407 (1903).

A Velocity Constant by a Conductimetric Method

Discussion

The saponification of ethyl acetate,

$$CH_3COOC_2H_5 + OH^- = CH_3COO^- + C_2H_5OH \tag{a}$$

proceeds by a second-order reaction. Hence if the initial concentrations of the reactants are both equal to a and if x is the number of moles per litre which have reacted after a time t, then

$$\frac{dx}{dt} = k(a-x)^2$$

Integration of this equation gives

$$k = \frac{1}{at} \cdot \frac{x}{a-x} \tag{1}$$

As indicated by reaction (a), the conductivity of the solution undergoes a marked decrease due to the replacement of the highly conducting hydroxyl ion by the acetate ion. The progress of this reaction may, therefore, be followed conductimetrically.[1] Equation (1) may be written in terms of conductivities rather than concentrations to give

$$k = \frac{1}{at} \cdot \frac{K_0 - K_t}{K_t - K_\infty} \tag{2}$$

where K_0 is the initial conductance of the solution, K_t the conductance of the solution after the time t and K_∞ the conductance when the reaction is complete. Rearrangement of equation (2) gives

$$K_t = \frac{1}{ak} \cdot \frac{K_0 - K_t}{t} + K_\infty$$

The velocity constant k can thus be calculated from a knowledge of the initial concentration of the reactants and a plot of K_t vs. $(K_0 - K_t)/t$.

Apparatus and Chemicals

Philips conductivity bridge, dip-type conductivity cell, thermostat, stop watch, ethyl acetate and sodium hydroxide.

Method

The conductivity cell is rinsed with distilled water. The cell is filled with a known volume of 0·02 M sodium hydroxide. A conical flask is filled with an equal volume of 0·02 M ethyl acetate. Both the flask and the cell are then immersed in a thermostat at 25°C until thermal equilibrium is attained. The ethyl acetate solution is then poured into the cell. Conductance readings should be made as soon as possible since the time of mixing marks the start of the reaction. The conductance should be measured at convenient intervals of time until the reaction is complete, i.e. after a period of approximately 1 hr, when a constant value for the conductance K_∞ is obtained.

A graph of K_t vs. $(K_0 - K_t)/t$ is then plotted, and the velocity constant k calculated from the gradient.

[1] J. WALKER, *Proc. Roy. Soc.* A, **78**, 157 (1906).

The Effect of Change of Temperature on the Rate of a Reaction

Discussion

The rates of most chemical reactions increase rapidly with increasing temperature. For simple reactions, the rate constant k is given, to a good approximation, by the Arrhenius equation

$$k = A\, e^{-E/RT} \tag{1}$$

where A and E are constants. The constant A is the frequency factor, and E the energy of activation. The logarithmic form of this equation is

$$\ln k = -E/RT + \ln A \tag{2}$$

Differentiation of equation (2) gives

$$d \ln k/dT = E/RT^2$$

which on integrating between the limits T and T' gives

$$\log k'/k = \frac{E}{2 \cdot 303R} \cdot \frac{T'-T}{T'T} \tag{3}$$

Hence, if the specific rate constants k and k' are determined at two temperatures, or if the ratio k'/k and T and T' are known, E may be calculated.

The reaction selected for study is the oxidation of hydrogen iodide by hydrogen peroxide in acid solution.

$$H_2O_2 + 2HI \rightarrow I_2 + 2H_2O$$

The reaction is first order with respect to each reactant.

Under the given experimental conditions the total volume of the reaction mixture is kept effectively constant.

By reducing the iodine formed with sodium thiosulphate and reacting the resultant sodium iodide with the acid already present, the hydrogen iodide concentration remains constant throughout the experiment.

Thus the reaction reduces to a pseudo first-order reaction. If experiments are carried out at two different temperatures, the ratio k'/k can be determined directly.

Apparatus and Chemicals

Pneumatic trough, 1 l. wide-mouth conical flask, stirrer, special burette as shown in Fig. 1, two stoppered test tubes, 20 vol. solution of hydrogen peroxide, concentrated sulphuric acid, 0·1 N sodium thiosulphate, starch solution, supply of ice.

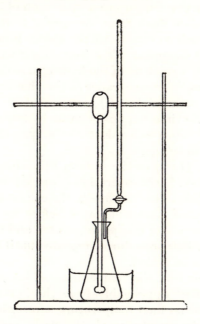

FIG. 1. Reaction flask with special burette.

Method

The 20 vol. hydrogen peroxide is diluted ten times with water. Two 25 ml portions of this freshly prepared 2 vol. hydrogen peroxide are measured with a pipette, added to two test tubes, and the test tubes placed in ice water.

Two grams of potassium iodide are dissolved in 500 ml of distilled water in the conical flask. The flask is placed in the ice bath. One part (10 ml) of concentrated sulphuric acid is mixed with two parts (20 ml) of distilled water, and this mixture added to the potassium iodide solution. The contents of the flask are allowed to cool until the temperature of the solutions is below 2°C.

The special burette is filled with 0·1 N sodium thiosulphate and the burette is placed so that its contents can be discharged into the wide-mouth conical flask. A few drops of starch solution are added to the hydrogen iodide solution in the conical flask and the contents of one of the test tubes poured into the flask. The hydrogen peroxide reacts with the hydrogen iodide, iodine is formed and a blue colour appears. Zero time (t_0) is the start of this reaction. This time is noted with a stop watch.

A measured volume of the 0·1 N sodium thiosulphate (1 ml is usually sufficient, but in the first case up to a maximum of 3 ml may be used), is run into the reaction flask from the special burette. If the sodium thiosulphate is present in excess all the iodine formed will react and the blue colour will disappear. This blue colour will shortly reappear due to the continuing oxidation of the hydrogen iodide by the hydrogen peroxide and the time of reappearance of the blue colour is noted. This time is t_1 and the interval ($t_1 - t_0$) is recorded in seconds. The time interval ($t_1 - t_0$) is the time required for the hydrogen peroxide to produce an amount of iodine equivalent to the quantity of sodium thiosulphate added at this stage.

A second portion (1 ml) of the sodium thiosulphate solution is run into the reaction flask. If this operation is carried out shortly after the appearance of the blue colour at time t_1, the colour will again disappear and a short interval of time will elapse before its reappearance. The time t_2 at which the colour reappears is noted and the interval ($t_2 - t_0$) in seconds recorded. This procedure is repeated until 10 ml of the sodium thiosulphate have been added.

At the end of this run, the temperature of the reaction mixture is noted, and if it differs appreciably from the initial value, the mean temperature and not its final temperature is recorded.

The experiment is repeated with the flask immersed in water at about 16°C (a higher temperature is undesirable as the time intervals would be too short for accurate observations). Again, a total of 10 ml of sodium thiosulphate is added in the same successive quantities as were added in the first experiment. The intervals of time for the second experiment, ($t_1' - t_0'$), ($t_2' - t_0'$), etc., are recorded in seconds.

The ratio of total times, e.g. ($t_i - t_0$)/($t_i' - t_0'$) are calculated, where $i = 1, 2, 3, ..., $. The mean value of these ratios is equal to the ratio of the velocity constants.

$$\left[\frac{(t_i - t_0)}{(t_i' - t_0')} \right]_{\text{mean}} = \frac{k'}{k}$$

The activation energy E is calculated from equation (3).

Determination of the Order of a Reaction

Discussion

The reaction between bromine and acetone is catalysed by acid in aqueous solution. The rate of the reaction dx/dt is given by

$$\frac{dx}{dt} = k[H^+]^a[Br_2]^b[CH_3COCH_3]^c \tag{1}$$

where a, b and c are the orders of the reaction with respect to the catalyst, bromine and acetone respectively. If the hydrogen ion concentration is kept constant in a series of experiments, equation (1) may be written

$$\frac{dx}{dt} = k'[Br_2]^b[CH_3COCH_3]^c \tag{2}$$

where $k' = k[H^+]^a$.

By investigating the kinetics of the reaction at constant hydrogen ion concentration the order of the reaction with respect to bromine and acetone may be determined.

Apparatus and Chemicals

Two 250 ml stoppered bottles, thermostat at 25°C, pipette, stop clock, 0·02 M bromine solution, 1 M acetone solution, 1 M sulphuric acid, 0·25 M potassium iodide solution, 0·01 N sodium thiosulphate solution and 1 M sodium bicarbonate solution.

Method

100 ml of approximately 0·02 M bromine solution is added to a 250 ml stoppered bottle and to another bottle a mixture of 50 ml M acetone and 50 ml M sulphuric acid is added. The bottles are suspended in a thermostat at 25°C and 15 min are allowed for the attainment of thermal equilibrium. The acetone solution is then added to the bromine solution and the stop clock is started at the same time. At 5 min intervals a 10 ml sample of the reaction mixture is withdrawn and allowed to run into a flask containing 10 ml of 1 M sodium bicarbonate solution and about 2 ml of 0·25 M potassium iodide solution.

The iodine liberated is titrated with 0·01 N sodium thiosulphate solution and a graph is plotted of titre against time. The reaction is followed until it is complete or for a maximum of 40 min.

The experiment is repeated with one bottle containing 100 ml of 0·02 M bromine again but with the other containing a mixture of 25 ml 1 M acetone, 25 ml of water and 50 ml of 1 M sulphuric acid.

A third experiment is performed using a mixture of 37·5 ml of 1 M acetone, 12·5 ml of water and 50 ml of 1 M sulphuric acid in the second bottle.

Assuming that the acetone is in all cases present in such large excess that its concentration may be assumed to be constant during the reaction, the order of the reaction with respect to bromine is deduced from the graphs. Another graph is plotted from the results to derive the order of the reaction with respect to acetone. The value of k' should also be derived.

Surface Tension using a Traube Stalagmometer

Discussion

The surface tension of a liquid is related to the weight of a drop of that liquid which falls freely from the end of a tube, by the expression

$$\gamma = \frac{mg}{2\pi r F}$$

where γ is the surface tension, m the mass of the drop, g the acceleration due to gravity, r the radius of the end of the tube and F is a function of $v/r^{1/3}$, where v is the volume of the drop.

When using the Traube stalagmometer to measure surface tension a fixed volume of liquid is delivered as freely falling drops from the end of a tube and the number of drops formed is counted. If the experiment is repeated with the same volume of a reference liquid, the unknown surface tension can be calculated from the relationship

$$\frac{\gamma_1}{\gamma_2} = \frac{m_1 g F_2}{m_2 g F_1} = \frac{v_1 d_1 F_2}{v_2 d_2 F_1}$$

where d_1 and d_2 are the densities of liquid 1 and liquid 2 respectively.

The function F varies only slightly over a large range of values of $v/r^{1/3}$ (see Experiment 34, Table 1), so that if the volumes of the drops of each of the liquids are not greatly different we may write

$$\frac{\gamma_1}{\gamma_2} = \frac{v_1 d_1}{v_2 d_2}$$

If V is the volume of each liquid delivered and if n_1 and n_2 are the number of drops of liquid 1 and liquid 2, respectively, then

$$v_1 = \frac{V}{n_1} \quad \text{and} \quad v_2 = \frac{V}{n_2}$$

hence

$$\frac{\gamma_1}{\gamma_2} = \frac{n_2 d_1}{n_1 d_2}$$

Apparatus and Chemicals

Traube stalagmometer (Fig. 1), receiving vessel, thermostat at 20°C, benzene and carbon tetrachloride.

Method

The scale on the stalagmometer is calibrated to find the number of scale divisions corresponding to one drop of liquid. This must be done for each liquid in turn before the main observations are commenced.

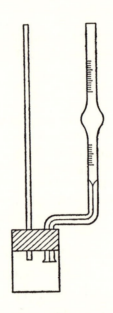

FIG. 1. A stalagmometer.

Care should be taken that the stalagmometer is thoroughly clean and dry before operations are begun. Having calibrated the scale, fractions of a drop can now be estimated, if necessary, with an accuracy of 0·05 of a drop.

The stalagmometer is filled with benzene, fitted into the receiving vessel and the assembly placed in a thermostat at 20°C so that the liquid in the stalagmometer is below the level of the thermostat liquid.

When the system has achieved thermal equilibrium the benzene is allowed to flow from the stalagmometer at a rate not greater than 15 drops per min. The rate of flow may be adjusted by attaching a piece of rubber tubing with a screw clip to the top of the stalagmometer.

The number of drops formed while the level of the benzene falls from the upper to the lower mark on the stalagmometer is counted. This determination is repeated until reproducible results are obtained, i.e. different determinations of drop number not varying by more than 0·5 of a drop.

The stalagmometer is cleaned and dried and the experiment repeated using a liquid of unknown surface tension, e.g. carbon tetrachloride.

From the results the surface tension of the liquid under investigation is calculated.

The surface tension of benzene at 20°C is 28·88 dyne cm^{-1}.

Interfacial Tension
(Micrometer Syringe Method)

Discussion

The size of a liquid drop issuing from a small orifice is governed by its surface tension, the value of which may be calculated from the equation

$$\gamma = \frac{mg}{2\pi r F}$$

where m is the mass of the drop, g the acceleration of gravity, r the external radius of the capillary tube and F is a factor[1] depending on $v/r^{1/3}$, where v is the volume of the drop. The product rF can be determined for two immiscible liquids A and B whose interfacial tension $_A\gamma_B$ is known. This factor can then be used to calculate the unknown interfacial tension $_A\gamma_C$ between one of the above liquids A and a third liquid C, providing the experiments with A and C are carried out under similar conditions to those for A and B.

This drop size technique is particularly useful in the determination of interfacial tensions, the micrometer syringe providing a convenient apparatus for the measurement of the volume of a drop. For two liquids, the interfacial tension is given by

$$_A\gamma_B = \frac{v(d_A - d_B)g}{2\pi r F}$$

where d_A and d_B are the densities of the liquids A and B ($d_A > d_B$).

Apparatus and Chemicals

Micrometer syringe, benzene, chloroform, carbon tetrachloride and chromic acid solution.

Method

The micrometer syringe is cleaned with chromic acid solution and then thoroughly washed with distilled water. Unless the tip of the syringe needle is exceptionally clean difficulty will be found in forming the drop so that it completely covers the tip. The tip of the needle may be completely wetted (r = external radius) or completely non-wetted (r = internal radius). If the chromic ions are not removed after cleaning the tip will be too easily wetted and if the organic phase is in the syringe it will be difficult to form a drop which completely covers the tip. On the other hand, if the needle is not cleaned with chromic acid and the aqueous phase is in the syringe the tube may be too hydrophobic and the drops of water may break off inside the needle. On the whole it is best to clean the syringe with chromic acid, wash it thoroughly and then perform the experiments so that the water is always inside the syringe.

When the syringe has been cleaned and washed, it is filled with distilled water and a drop is then slowly formed in benzene. The delivery of the last 10 per cent of the drop should take at least $1\frac{1}{2}$ min, the correct drop volume v is only obtained by using extreme care in the manipulation of the micrometer screw. A number of droplets are formed and their average volume recorded. The densities of benzene, water and the interfacial tension $_{H_2O}\gamma_{C_6H_6}$ at the temperature of the experiment are then obtained. Hence a value of g/rF can be obtained which can be combined with further experimental results and further data from the tables below to calculate the interfacial tension of water/chloroform $_{H_2O}\gamma_{CHCl_3}$ and water/carbon tetrachloride $_{H_2O}\gamma_{CCl_4}$.

TABLE 1.[1] VARIATION OF F WITH $v/r^{1/3}$

$r/v^{1/3}$	F	$r/v^{1/3}$	F
0·00	1·0000	1·00	0·6098
0·30	0·7256	1·05	0·6179
0·35	0·7011	1·10	0·6280
0·40	0·6828	1·15	0·6407
0·45	0·6669	1·20	0·6535
0·50	0·6515	1·225	0·656
0·55	0·6362	1·25	0·652
0·60	0·6250	1·30	0·640
0·65	0·6171	1·35	0·623
0·70	0·6093	1·40	0·603
0·75	0·6032	1·45	0·583
0·80	0·6000	1·50	0·567
0·85	0·5992	1·55	0·551
0·90	0·5998	1·60	0·535
0·95	0·6034		

[1] W. D. HARKINS and F. E. BROWN, *J. Amer. Chem. Soc.*, **41**, 499 (1919).

Variation of the Surface Tension of a Liquid with Temperature

Discussion

The surface tension of a liquid varies with the temperature. A number of equations have been suggested for the relationship between temperature and surface tension, one of the most satisfactory being

$$\gamma\left(\frac{M}{d}\right)^{2/3} = k(t_c - 6 - t) \tag{1}$$

where d is the density, M the molecular weight, t_c the critical temperature of the liquid under investigation, and k is a constant.

For many substances, e.g. carbon disulphide, carbon tetrachloride and diethyl ether the temperature coefficient k is constant and has a value of about 2·1. For certain liquids such as water, alcohols and carboxylic acids k is less than 2·1 and varies with temperature. This behaviour is attributed to the fact that these liquids are associated and the effective molecular weight is not M but Mx where x is the degree of association. The temperature coefficient is thus given by

$$-\frac{d[\gamma(Mx/d)^{2/3}]}{dT} = k = 2\cdot1 \tag{2}$$

Thus if k' is the experimental value based on the assumption that $x = 1$, then

$$x = (2\cdot1/k')^{3/2} \tag{3}$$

The surface tension of a liquid may be determined from observations of the differential capillary rise obtained when two capillaries of differing bore are immersed in the liquid, see Fig. 1. The ratio of the surface tensions for two liquids A and B is given by the equation

$$\frac{\gamma_A}{\gamma_B} = \frac{(d_A \cdot \Delta h_A)}{(d_B \cdot \Delta h_B)} \tag{4}$$

where Δh_A and Δh_B are the differences in the levels of liquids A and B, respectively, in the two capillaries. Knowing Δh_A and Δh_B, the densities of the two liquids and the surface tension of one of the liquids,[1] the surface tension of the other liquid can be determined. If the experiment is carried out at various temperatures, then according to equation (1) a plot of $\gamma(M/d)^{2/3}$ vs. t should be linear.

The surface tension is related to the capillary rise by

$$\gamma = \frac{rhdg}{2\cos\theta} \qquad (5)$$

where r is the radius of the capillary, g the acceleration due to gravity and θ is the contact angle. If the capillary tubes are completely wetted, $\theta = 0$ and for two capillary tubes of radii r_1 and r_2

$$\Delta h = \gamma\left(\frac{2}{dg}\right)\left(\frac{1}{r_1} - \frac{1}{r_2}\right) \qquad (6)$$

Using a liquid of known surface tension it is thus possible to arrive at a value for

$$\left(\frac{1}{r_1} - \frac{1}{r_2}\right)$$

If two standard bore capillaries are provided, values of the radii are known and it is unnecessary to use a liquid of known surface tension to calibrate the system.

Apparatus and Chemicals

Two differential capillary tubes of internal bore diameters 1·5 to 2 mm and 0·3 to 0·4 mm, glass-sided thermostat, cathetometer, benzene.

Method

The boiling tube and the capillaries are thoroughly cleaned with chromic acid solution, water and alcohol. The apparatus is dried and a small volume of distilled water placed in the test tube and the capillary tubes inserted as shown in Fig. 1. Using a cathetometer the differential capillary rises, Δh, are measured for distilled water at 20°C, 30°C and 50°C. The boiling tube and capillaries are again cleaned and dried and the experiment repeated at the above three temperatures with benzene.

Thermal equilibrium must be attained at each working temperature, the temperature of the liquid in the boiling tube should be checked to ensure that it is identical with that of the thermostat. This equilibration would normally occupy a period of 15 to 20 min. A slight pressure should be applied to the system, see Fig. 1, and oscillation of the menisci obtained. On removal of the pressure, the differential level Δh should not have changed. If a change is observed the tubes are obviously dirty, and should be cleaned again.

D

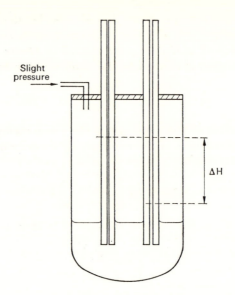

FIG. 1. Differential capillary rise apparatus.

The densities of the liquids should be determined by means of a pyknometer. The surface tension of benzene at the three temperatures should be calculated using equation (4) if the bores of the capillaries provided are unknown. If the bores are known, equation (5) may be used. A graph of $\gamma(M/d)^{2/3}$ vs. t is plotted and the slope k found. A value of x for water should be calculated from equation (3).

[1] *Handbook of Chemistry and Physics* (edited by C. D. HODGMAN), Chemical Rubber Publishing Co.

Adsorption Isotherm

Discussion

The variation of the extent of adsorption of a substance with concentration may frequently be represented by the classical adsorption isotherm, sometimes known as the Freundlich adsorption isotherm.

$$a = kc^{1/n} \qquad (1)$$

where a is the mass of substance adsorbed per unit mass of adsorbent, c the equilibrium concentration of the substance adsorbed and both k and n are constants.

Equation (1) may be written

$$\log a = \log k + 1/n \log c \qquad (2)$$

Thus, if $\log a$ is plotted against $\log c$ a straight line of slope $1/n$ and intercept $\log k$ should be obtained.

When adsorption is limited to a monolayer the behaviour of the system is very well represented by another isotherm known as the Langmuir isotherm.

$$a = \frac{k_1 c}{1 + k_2 c} \qquad (3)$$

where k_1 and k_2 are constants. This equation may be written

$$c/a = (k_2/k_1)c + 1/k_1 \qquad (4)$$

and the isotherm may be tested by plotting c/a against c.

Apparatus and Chemicals

(a) Adsorption of acetone on charcoal: iodine flask, charcoal, acetone, 2 N sulphuric acid, 2 N sodium hydroxide, 0·1 N iodine solution and 0·1 N sodium thiosulphate solution.

(b) Adsorption of acetic acid on charcoal: N acetic acid, 0·1 N sodium hydroxide and charcoal.

Method

The adsorption of either acetic acid or acetone on charcoal may be investigated.

(a) *Adsorption of acetone on charcoal*

An approximately 0·2 M solution of acetone is prepared by mixing 15 ml of acetone with 1 l. of water. Fifty millillitres of this solution is then diluted to 250 ml in a graduated flask.

The acetone is determined in the following manner, 10 ml of the acetone solution is transferred to an iodine flask and about 12 ml of 2 N sodium hydroxide is added, together with about 12 ml of water. Forty millilitres of standard 0·1 N iodine solution is added, the flask stoppered, well shaken and allowed to stand for 10 min. About 12 ml of 2 N sulphuric acid and 12 ml of water is added to the flask and the iodine liberated is titrated with standard 0·1 N sodium thiosulphate solution. One millilitre of 0·1 N iodine *consumed* is equivalent to 0·000967 g acetone.

A hundred millilitres of the 0·2 M acetone solution is now shaken with 1·00 g of charcoal for 2 min. The mixture is filtered through a dry fluted filter paper, the first 10 ml of the filtrate (from which adsorption may have occurred on the filter paper) being rejected. The acetone in the filtrate is then determined volumetrically after dilution as described previously.

The experiment is repeated using, successively, 0·05, 0·02 and 0·005 M acetone solutions obtained by suitable dilution of the original 0·2 M solution. The filtrates from these experiments do not need to be diluted before being analysed, and 20 ml of decinormal iodine solution gives a sufficient excess.

The acetone adsorbed, a, in each experiment is calculated by difference. A plot of log a against log c is made where c is the concentration of acetone found in the filtrates. The values of the constants k and n are determined from the intercept and slope, respectively.

(b) *Adsorption of acetic acid on charcoal*

The normal acetic acid solution is diluted by factors of 2, 4, 10, 20 and 50 to a total volume of 100 ml of solution in each case. 100 ml of N acetic acid is placed in the first of six conical flasks and the 100 ml quantities of the diluted solutions are placed in the remaining five flasks. A weighed sample (approximately 3 g) of activated charcoal is added to each flask and the solutions are agitated at frequent intervals during a 15 min period. The contents of each flask are filtered through a *dry* filter paper. The first 10 ml of filtrate is rejected. A known volume of each filtrate is titrated with 0·1 N sodium hydroxide. Suitable volumes are 2 ml from the most concentrated solution, 5 ml from the next, 10 ml from the next, 25 ml from the next and 50 ml from each of the two most dilute solutions.

It is important that the time between adding the charcoal and filtering the solution is the same in each case. Furthermore, the period of 15 min allowed for the attainment of equilibrium is a minimum and a period of 30 min or even an hour would be preferable.

The amount of acetic acid adsorbed per gram of adsorbent is calculated and the Freundlich and Langmuir isotherms are tested by plotting log a vs. log c and c/a vs. c respectively.

Variation of Conductance with Concentration (Strong and Weak Electrolytes)

Discussion

The specific conductance κ of any conductor is defined as the reciprocal of the specific resistance, that is

$$\kappa = \frac{l}{aR}\Omega^{-1}\ cm^{-1}$$

where R is the resistance, l is the length in centimetres and a is the cross-sectional area in square centimetres of the conductor. The equivalent conductance Λ is obtained by multiplying the specific conductance by the volume of the solution measured in cubic centimetres containing 1 g equivalent of the electrolyte

$$\Lambda = \kappa \cdot v\ \Omega^{-1}\ cm^2\ eqt.^{-1} \tag{1}$$

Hence, if c is the concentration of the solution in gram equivalents per litre, v is equal to $1000/c$, neglecting the difference between millilitre and cubic centimetre, and

$$\Lambda = \kappa \cdot 1000/c\ \Omega^{-1}\ cm^2\ eqt.^{-1} \tag{2}$$

Note that an equivalent is the amount of substance associated with the transfer of one faraday of electricity. Hence, if one mole carries n faradays of electricity there will be n equivalents in one mole. Only in the case where one mole is associated with the transfer of one faraday are the mole and equivalent identical.

Both specific and eqivalent conductance of a solution vary with concentration. Specific conductance increases with increase in concentration. Equivalent conductance increases with dilution, a limiting value Λ_0 being obtained at infinite dilution (zero concentration). The equivalent conductance at zero concentration of an electrolyte is not the same as the equivalent conductance of the pure solvent. It is the equivalent conductance of the solution when the ions of the electrolyte are on infinite distance apart, and hence free from interaction.

For certain electrolytes, Kohlrausch suggested the empirical relationship

$$\Lambda = \Lambda_0 - b\sqrt{c} \tag{3}$$

where b is a constant. These electrolytes are known as strong electrolytes and a plot of Λ vs. $\sqrt{c}$ is linear for concentrations up to 0·01 N. The equation is accurate to within a few per cent, however, for concentrations of up to 0·1 N in the case of uni-univalent electrolytes. Dilute aqueous solutions of many other electrolytes, e.g. acetic acid, do not give a linear graph. For these electrolytes, known as weak electrolytes, the value of the equivalent conductance increases rapidly as $\sqrt{c}$ approaches zero, and the graph cannot be accurately extrapolated to obtain a value of Λ_0.

Apparatus and Chemicals

Doran conductance bridge, conductance cell (cell constant about 0·36), thermostat, 0·1 N or 0·01 N potassium chloride solution, 0·1 N sodium chloride solution, 0·1 N acetic acid and conductivity water. (The conductance of the conductivity water should be measured separately and subtracted from the observed conductance of the solutions.)

Conductivity cells should be carefully washed and steamed before use. They are fragile and expensive and should be handled with care. The electrodes should be coated with a thin film of platinum black, consequently they must not be handled and once inserted in the cell should not be moved during the experiment. The electrodes should be kept in distilled water when not in use.

Conductivity solutions should be made up and diluted by weight. If it is not possible to adhere strictly to this procedure, all dilutions must be performed with great care. A 0·1 D, or 0·01 D solution of potassium chloride may be used to calibrate the conductivity cell. Demal solutions are prepared by dissolving 1 g equivalent of the electrolyte in 1000 g of water. A sufficiently accurate solution may be prepared by dissolving a weighed quantity of potassium chloride in conductivity water contained in a graduated flask, and adjusting to volume after bringing the temperature up to 25°C. The conductivity can then be calculated from the empirical equation due to Shedlovsky.[4]

$$\Lambda = 149{\cdot}82 - 93{\cdot}85\sqrt{(c)} + 94{\cdot}9c(1 - 0{\cdot}2274\sqrt{c})$$

where c is the concentration in g equivalents per litre.

Demal solutions differ only slightly from normal solutions. For potassium chloride the specific conductivities in ohm^{-1} cm^{-1} are

Temperature	0·1 D	0·1 N	0·01 D	0·01 N
18 °C	0·011167	0·01120	0·0012205	0·001224
25 °C	0·012856	0·01289	0·0014088	0·001412

Ordinary tap water must not be used for conductivity work. For most purposes, except those demanding the highest accuracy, water having a specific conductance of $10^{-6} \, \Omega^{-1} \, cm^{-1}$ is sufficiently pure. A convenient method for the preparation of "conductivity water" is to pass ordinary distilled water through a column of mixed ion-exchange resins. Conductivity water free from dissolved organic material may be obtained from special stills, e.g. Baird & Tatlock double unit Baraglass still.

Method

Using a pipette, a measured volume of standard potassium chloride solution is added to the conductivity cell. The electrodes are washed with distilled water, dried by touching with filter paper, immersed in potassium chloride solution, and then placed in the conductivity cell. The electrodes should be at least 1 cm below the surface of the solution.

The conductance cell is supported in a thermostat at 25°C. A period of at least 10 min should be allowed to elapse to ensure that the system attains thermal equilibrium. The cell is then connected to the conductance bridge and the resistance of the solution is measured.

The cell constant C is given by the equation

$$C = R\kappa$$

where R is the resistance and κ the specific conductance of the solution of potassium chloride.

The experiment is repeated with a 0·1 N solution of sodium chloride using the same amount of solution in the conductivity cell. The solution is diluted two-fold and the resistance of the diluted solution measured. This procedure is repeated making seven two-fold dilutions. In each case the specific conductance is calculated from the equation

$$\kappa = C/R$$

where R is the resistance of the sodium chloride solution.

The equivalent conductance is calculated from equation (2). The values found for the equivalent conductance are plotted against $\sqrt{c}$. The slope of this graph is determined.

The experiment is repeated with the 0·1 N acetic acid solution and a graph of equivalent conductance versus $\sqrt{c}$ drawn. The shape of this graph is compared with that obtained for sodium chloride.

Conductimetric Titration of an Acid Mixture

Discussion

On the addition of a solution of a strong base to that of a strong acid, there is initially a decrease in conductivity due to the replacement of the fast moving hydrogen ion by the slower alkali metal ion. The conductivity subsequently increases again with the increase in concentration of hydroxyl ions when the titration is complete. Hence if the conductivity is plotted against the volume of base solution added, the end point of the titration is given by a sharp inflection in the graph.

If a strong base, however, is added to a solution of a weak acid, there is an initial gradual increase in the conductivity as the feebly dissociated weak electrolyte is converted into its salt, a stronger electrolyte. Subsequently at the end point a much more rapid conductivity increase occurs due to the presence of excess hydroxyl ions.

If a mixture of a strong acid and a weak acid is titrated by a strong base, an initial decrease in conductivity is followed by an increase and then by a more rapid increase, thereby giving two end points, see Fig. 1.

Apparatus and Chemicals

Doran conductivity bridge, dip-type conductivity cell, a mixture of $0 \cdot 1$ M hydrochloric acid and $0 \cdot 1$ M acetic acid, and $0 \cdot 1$ M sodium hydroxide solution.

Method

The cell vessel is cleaned and dried. The electrodes are rinsed with distilled water and then with the given acid mixture. The preparation is completed by removal of excess acid with filter paper. The electrodes are immersed in a beaker containing 50 ml of the acid mixture and then connected to the conductivity bridge. The conductivity of the solution is then measured. The $0 \cdot 1$ N caustic soda solution is then run into the acid solution and the conductance of the solution is measured after the addition of 4, 8, 12, 16, 20, 29, 33, 37, 41, 45, 54, 58, 62, 66

D*

and 70 ml of caustic soda. Before the measurement of the conductance is made the solution should be thoroughly stirred.

The total volume of alkali and acid, v, is determined for each measured conductivity value. A graph of vK vs. burette readings is then plotted and three straight lines are drawn through the points in order to obtain the two equivalence points.

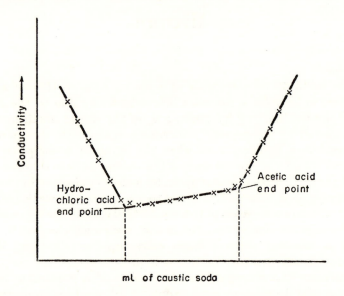

Fig. 1. Variation of conductivity with volume of added caustic soda solution.

Solubility by a Conductimetric Method

Discussion

In a saturated solution of a sparingly soluble salt such as lead sulphate the solution is so dilute that its equivalent conductivity may be assumed to be equal to its equivalent conductivity at infinite dilution. The specific conductivity of the lead sulphate may be obtained by determining the specific conductivity of the saturated solution and then subtracting the specific conductivity due to the solvent.

$$\kappa_{solution} - \kappa_{solvent} = \kappa_{PbSO_4}$$

The equivalent conductivity of the lead sulphate is given by

$$\Lambda_{PbSO_4} = \kappa_{PbSO_4} v = \frac{\kappa_{PbSO_4}}{c}$$

where v is the volume in millilitres containing one equivalent of solute and c is the concentration of the solution in gram equivalents of solute per millilitre.

Assuming that the equivalent conductivity of the saturated solution is approximately equal to the equivalent conductivity at infinite dilution we may write

$$\lambda_{Pb^{2+}} + \lambda_{SO_4^{2-}} = \frac{\kappa_{PbSO_4}}{c}$$

or

$$c = \frac{\kappa_{PbSO_4}}{\lambda_{Pb^{2+}} + \lambda_{SO_4^{2-}}}$$

where $\lambda_{Pb^{2+}}$ and $\lambda_{SO_4^{2-}}$ are the ionic conductivities at infinite dilution of the lead and sulphate ions, respectively.

Apparatus and Chemicals

Cambridge conductivity bridge, conductivity cell, stirrer, conductivity water and lead sulphate.

Method

The conductivity cell is rinsed out thoroughly with conductivity water and then filled with this water making sure that no air bubbles adhere to the surfaces of the electrodes. The conductivity of the water is measured. The cell is rinsed out, refilled with conductivity water and the conductivity is again measured. This process is repeated until reproducible results are obtained. All determinations should be carried out at 25°C.

A sample of the lead sulphate is taken and shaken repeatedly with conductivity water to remove any soluble impurities. The washed sample of lead sulphate is suspended in conductivity water in a flask which has previously been thoroughly cleaned with conductivity water. The suspension is warmed to about 30°–35°C, placed in a thermostat operating at 25°C and stirred vigorously until equilibrium has been established. The specific conductivity of this solution is determined by repeated measurements on fresh samples until reproducible results are obtained. Samples should be withdrawn through a glass wool filter to prevent solid particles entering the conductivity cell.

Given the following equivalent ionic conductivities at infinite dilution the solubility of lead sulphate at 25°C in grams per 100 g of water can be calculated.

$$Pb^{2+} \qquad 72 \cdot 0 \ \Omega^{-1} \ cm^2 \ eqt.^{-1}$$
$$SO_4^{2-} \qquad 79 \cdot 0 \ \Omega^{-1} \ cm^2 \ eqt.^{-1}$$

Transport Numbers (Hittorf's Method)

Discussion

In electrolysis all the ions in solution share in carrying the current and the fraction of the total current carried by a particular ion is known as the transport number of that ion. The sum of the transport numbers of all the ions in a solution will thus be equal to unity.

Current is carried by virtue of the migration of the ions between the electrodes and if the number of equivalents of a particular ion that have migrated during electrolysis can be determined and expressed as a fraction of the total amount of electricity passed the resulting figure will be the transport number of that ion. This is the basis of the Hittorf method for determining transport numbers.

In this experiment the electrolysis of 0·05 M copper sulphate is carried out between copper electrodes. The total amount of electricity passed during electrolysis is found from the increase in weight of the cathode of a copper coulometer which is included in the circuit. The increase in the amount of copper in solution around the anode of the Hittorf cell is deduced from an analysis of the anolyte at the conclusion of electrolysis. From a knowledge of the total amount of electricity passed and the increase in the number of copper ions in the anolyte, the amount of copper which has migrated and hence the transport number of the copper ion may be calculated.

Apparatus and Chemicals

Hittorf cell, copper coulometer, d.c. supply (about 60 V) milliammeter, control resistance, 0·05 M copper sulphate solution, 0·1 N sodium thiosulphate solution, potassium iodide and copper sulphate solution for use in the coulometer.

Method

About 20 ml of the 0·05 M copper sulphate solution is added to a weighed beaker and the beaker and contents are weighed again. The copper content of the solution is determined with potassium iodide and 0·1 N sodium thiosulphate solution, the result being expressed as equivalents of copper per gram of solvent.

The cathode of the copper coulometer is cleaned with distilled water and alcohol, dried in an oven and finally weighed. The cathode is inserted into the coulometer and the vessel is filled with the solution provided. The Hittorf cell is filled with 0·05 M copper sulphate solution having been first rinsed with the solution, the copper electrodes are inserted and the circuit is connected up as shown in Fig. 1.

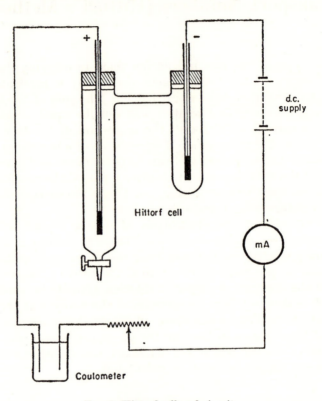

FIG. 1. Hittorf cell and circuit.

The d.c. supply is switched on and the current is adjusted to about 10 mA. Electrolysis is allowed to proceed for about $2\frac{1}{2}$ hr.

At the conclusion of electrolysis about two-thirds of the anolyte is run off into a previously weighed beaker. The copper content of the solution is determined, the result being expressed as "x grams of water contain y equivalents of copper" where x is the weight of water in the sample of anolyte titrated. This result is combined with the previous analysis and the increase in the number of equivalents of copper in x grams of water is calculated.

The cathode is withdrawn from the coulometer, washed with distilled water and alcohol, dried and weighed as before.

Let the increase in the weight of the coulometer cathode be a equivalents of copper. This is also the amount of copper which will have dissolved from the anode in the Hittorf cell giving rise to an increase in the amount of copper in solution. Some of this copper will have migrated under the influence of the electric field so that the observed increase in the amount of copper in the anolyte will be less than a equivalents.

Let the observed increase of copper in the anolyte be b equivalents of copper. Then the amount of copper which has migrated is $(a-b)$ equivalents. Now $(a-b)$ will be proportional to the amount of current carried by the copper ions whilst a will be proportional to the total amount of electricity passed. The transport number of the copper ion $n_{Cu^{2+}}$ will thus be given by

$$n_{Cu^{2+}} = \frac{a-b}{a}$$

As $n_{Cu^{2+}} + n_{SO_4^{2-}} = 1$ the transport number of the sulphate ion is given by

$$n_{SO_4^{2}} = 1 - n_{Cu^{2+}}$$

Comment should be made on the value of the result in view of the experimental errors involved.

Dissociation Constant of a Weak Acid (Approximate e.m.f. Method)

Discussion

The dissociation constant of a weak acid may be accurately determined from e.m.f. measurements using a cell without a liquid junction. A potentiometric method involving a cell with a liquid junction gives an approximate result, but is more widely used, as it provides a ready and convenient method. The method involves the titration of the weak acid with a strong base, the pH change during the titration being followed by observing the potential of a glass electrode.

Consider a solution which originally contained a moles of a weak monobasic acid HA, to which b moles of a strong monacidic base, MOH, have been added, b being less than a. The reaction of the acid and the base may be represented

$$HA + M^+ + OH \rightarrow M^+ + A^- + H_2O$$

Some of the remaining acid will be dissociated

$$HA + H_2O \rightleftharpoons H_3O^+ + A^-$$

and some of the salt will undergo hydrolysis

$$A^- + H_2O \rightleftharpoons HA + OH^-$$

The solution will contain undissociated HA molecules, H^+ ions, OH^- ions, A^- ions and M^+ ions. As the solution must be electrically neutral, the sum of all the positive charges must be equal to that of all the negative charges, and denoting the number of moles of each ion present in the solution as n with an appropriate subscript

$$n_{M^+} + n_{H^+} = n_{A^-} + n_{OH^-}$$

The number of moles of M^+ ions must be equal to the number of moles of base added, hence

$$b + n_{H^+} = n_{a^-} + n_{OH^-} \tag{1}$$

The A^- ions which exist in the solution must have come from the acid originally present, so that

$$a = n_{HA} + n_{A^-} \tag{2}$$

where n_{HA} is the number of moles of undissociated acid existing in the mixture. The dissociation constant of the acid is given by

$$K_a = \frac{a_{H^+} \cdot a_{A^-}}{a_{HA}}$$

$$= a_{H^+} \cdot \frac{m_{A^-}}{m_{HA}} \cdot \frac{\gamma_{A^-}}{\gamma_{HA}} \tag{3}$$

Suppose now that the weight of solvent in the solution is w kilograms, then $n_{A^-} = wm_{A^-}$ and $n_{HA} = wm_{HA}$. Multiplying the numerator and denominator of equation (3) by w gives

$$K_a = a_{H^+} \cdot \frac{wm_{A^-}}{wm_{HA}} \cdot \frac{\gamma_{A^-}}{\gamma_{HA}}$$

$$= a_{H^+} \cdot \frac{n_{A^-}}{n_{HA}} \cdot \frac{\gamma_{A^-}}{\gamma_{HA}} \tag{4}$$

From equation (1)

$$n_{A^-} = b + n_{H^+} - n_{OH^-}$$

and from equation (2)

$$n_{HA} = a - n_{A^-}$$
$$= a - b - (n_{H^+} - n_{OH^-})$$

Substituting in equation (4)

$$K_a = a_{H^+} \cdot \frac{b + n_{H^+} - n_{OH^-}}{a - b - (n_{H^+} - n_{OH^-})} \cdot \frac{\gamma_{A^-}}{\gamma_{HA}}$$

or

$$a_{H^+} = K_a \cdot \frac{a - b - (n_{H^+} - n_{OH^-})}{b + n_{H^+} - n_{OH^-}} \cdot \frac{\gamma_{HA}}{\gamma_{A^-}} \tag{5}$$

Remembering that $n_{H^+} = wm_{H^+}$ and $n_{OH^-} = wm_{OH^-}$, equation (5) may be written

$$a_{H^+} = K_a \cdot \frac{a - b - w(m_{H^+} - m_{OH^-})}{b + w(m_{H^+} - m_{OH^-})} \cdot \frac{\gamma_{HA}}{\gamma_{A^-}} \tag{6}$$

Also, as $m_{H^+} \cdot m_{OH^-} = 10^{-14}$, the ionic product for water, it may be understood that if m_{H^+} is between 10^{-4} and 10^{-10}, the term $(m_{H^+} - m_{OH^-})$ becomes negligibly small and equation (6) reduces to

$$a_{H^+} = K_a \cdot \frac{a - b}{a} \cdot \frac{\gamma_{HA}}{\gamma_{A^-}}$$

or, taking logarithms

$$pH = pK_a + \log\frac{a}{a-b} + \log\frac{\gamma_{A^-}}{\gamma_{HA}}$$

Moreover, if the solution is sufficiently dilute for the activity coefficient terms to be neglected

$$pH = pK_a + \log\frac{a}{a-b}$$

At the point of half neutralization where $b = a/2$

$$pH = pK_a \qquad\qquad (7)$$

This point of half neutralization can be determined from a plot of pH vs. v ml of base added, but it is more accurate to plot $\Delta pH/\Delta v$ vs. v ml. Both graphs should be drawn. The first indicates the region of neutralization and gives the value of the pH at the point of half neutralization. The second differential graph gives the neutralization point as a peak.

Apparatus and Chemicals

Cambridge bench pH meter, glass electrode, calomel electrode, burette, beaker, 0·1 N sodium hydroxide and M/64 benzoic acid.

Method

Before proceeding with the experiment, the manufacturer's manual for the use of the pH meter should be consulted. The pH meter is standardized using two buffer solutions of known pH.

The electrodes are rinsed well with distilled water. The solution of the weak acid (50 ml) is added to the beaker. The electrodes are immersed in the solution. The standard alkali solution is placed in the burette, and the burette is arranged to deliver into the beaker. The base is added in small amounts, the volume is adjusted from time to time to give evenly spaced points on the neutralization curve. The critical regions are those of complete neutralization and half neutralization. A small inaccuracy in the pH measurements gives a large inaccuracy in the value of the dissociation constant. After each addition of base, the burette jet is kept away from the solution during pH measurements.

The solution is stirred well and a period of about 1 min after each addition of the base is allowed to elapse before the pH is read. The titration is continued until the pH no longer alters with the addition of further amounts of base.

From the data obtained, plots of (a) pH vs. v ml of base added, and (b) $\Delta \text{pH}/\Delta v$ vs. m ml of base added, should be drawn. From graph (b) the number of millilitres of base required for neutralization is determined. The pH at the point of half neutralization is read from graph (a). The $\text{p}K_a$ is given by equation (3).

Dissociation Constant of a Weak Acid (Conductimetric Method)

Discussion

According to the so-called Ostwald dilution law, for a binary weak electrolyte, e.g. the weak acid HA,

$$K_c = \frac{\alpha^2 \cdot c}{(1-\alpha)}$$

where α is the degree of dissociation, c is the concentration, and K_c is a constant sometimes referred to as the "classical" dissociation constant. According to the Arrhenius dissociation theory, the degree of dissociation is given by the ratio of the equivalent conductance to the conductance at infinite dilution

$$\alpha = \Lambda/\Lambda_0$$

Hence, substituting for α in the expression for the dissociation constant we have

$$K_c = \frac{\Lambda^2 \cdot c}{\Lambda_0(\Lambda_0 - \Lambda)}$$

$$\frac{\Lambda}{\Lambda_0^2} = K_c \frac{(1 - \Lambda/\Lambda_0)}{\Lambda c}$$

$$\Lambda \cdot c = K_c\left(\frac{\Lambda_0^2}{\Lambda} - \Lambda_0\right) \tag{1}$$

A plot of $\Lambda \cdot c$ vs. $1/\Lambda$ will thus give a straight line as shown in Fig. 1. At $c = 0, OX = 1/\Lambda_0$, that is $\Lambda_0 = 1/(OX)$. At $1/\Lambda = 0, OY = -K_c \cdot \Lambda_0$, therefore

$$(OX)(OY) = -K_c \tag{2}$$

where K_c is the "classical" dissociation constant.

Apparatus and Chemicals

Doran conductance bridge, conductivity cell suitable for a weak electrolyte, 25 ml pipette, 50 ml graduated flask, conductivity water, 0·01 D potassium chloride, 0.4×10^{-3} M, 2:4:6 trimethyl benzoic acid.

102

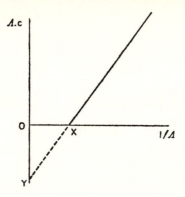

FIG. 1. Plot of $\Lambda \cdot c$ vs. $1/\Lambda$.

Method

The conductivity cell is rinsed three times with the solution of potassium chloride. The cell constant is determined.

The cell is emptied, rinsed with conductivity water and subsequently rinsed three times with the acid solution. A measured amount of the acid solution is added to the conductivity cell. Sufficient solution should be added to enable the electrodes to be immersed at least 1 cm below the surface of the solution. The conductance of the solution is determined.

Solutions of 2:4:6 trimethyl benzoic acid ($0 \cdot 35 \times 10^{-3}$ M, $0 \cdot 3 \times 10^{-3}$ M, $0 \cdot 25 \times 10^{-3}$ M, $0 \cdot 2 \times 10^{-3}$ M) are prepared by dilution with conductivity water and the conductance determined at each dilution.

A plot of Λc vs. $1/\Lambda$ is prepared (see Fig. 1), and hence the value of the "classical" dissociation constant K_c is calculated by equation (2).

Hydrolysis Constant by a Conductimetric Method

Discussion

When the salt of a strong acid and a weak base such as anilinium chloride is dissolved in water, hydrolysis occurs, the cations of the salt interacting with the solvent to form the conjugate base of the salt and the conjugate acid of the solvent.

$$C_6H_5NH_3^+ + H_2O \rightleftharpoons C_6H_5NH_2 + H_3O^+$$
$$\underset{1-\alpha}{} \qquad\qquad \underset{\alpha}{} \quad \underset{\alpha}{}$$

If a fraction α of the salt hydrolyses, the hydrolysis constant is given by

$$K_h = \frac{\alpha^2}{(1-\alpha)v} \tag{1}$$

where v is the volume in litres containing one mole of salt.

The conductivity of a solution of the salt in which hydrolysis has occurred will be made up partly of the conductivity due to the ions of the salt and partly of the conductivity due to the ions produced by hydrolysis. If the degree of hydrolysis is α and Λ_v is the equivalent conductivity of the solution at dilution v, then

$$\Lambda_v = (1-\alpha)\Lambda_v' + \alpha\Lambda_v'' \tag{2}$$

where Λ_v' is the equivalent conductivity of a solution of the un-hydrolysed salt at dilution v and Λ_v'' is the equivalent conductivity of hydrochloric acid at dilution v.

The first term on the right-hand side of equation (2) is the conductivity due to the unhydrolysed salt ions and the second term is the conductivity of the hydrochloric acid produced by the hydrolysis. The conductivity of the free aniline may be considered to be negligible.

Rearranging equation (2)

$$\alpha = \frac{\Lambda_v - \Lambda_v'}{\Lambda_v'' - \Lambda_v'} \tag{3}$$

Λ_v is calculated from conductivity measurements on solutions of anilinium chloride and Λ_v' is obtained from the conductivity of solutions in which hydrolysis has been almost completely suppressed by the addition of excess aniline. Knowing values of Λ_v'' for hydrochloric acid at various dilutions it is possible to calculate α from equation (3) and hence to calculate K_h from equation (1). Having calculated K_h and knowing the value of the ionic product for water K_w, it is possible to calculate the dissociation constant of aniline K_b from the relationship

$$K_h = \frac{K_w}{K_b} \tag{4}$$

Apparatus and Chemicals

Cambridge conductivity bridge, conductivity cell, aniline and anilinium chloride.

Method

An N/32 solution of anilinium chloride in water is prepared. Twenty millilitres of this solution is placed in the conductivity cell and the conductivity at 25°C is determined. The solution is diluted to N/64 and N/128 and the conductivity is determined at each dilution.

An N/32 solution of anilinium chloride in N/32 aniline is now prepared and the conductivity of this solution is determined at 25°C. The solution is diluted with N/32 aniline so that solutions which are N/64 and N/128 with respect to anilinium chloride are obtained. The conductivities of these solutions are then determined.

The cell constant of the conductivity cell used in the experiment is also determined.

Given the data below the results are used to calculate: (a) the degree of hydrolysis at each concentration; (b) the hydrolysis constant of anilinium chloride; (c) the dissociation constant of aniline.

Λ_v'' for hydrochloric acid at 25°C

N/32	393 Ω^{-1} cm^2 eqt.$^{-1}$
N/64	399 Ω^{-1} cm^2 eqt.$^{-1}$
N/128	401 Ω^{-1} cm^2 eqt.$^{-1}$

K_w for water at 25°C $= 1.01 \times 10^{-14}$ (g ions)^{2}l.$^{-2}$.

The Standard Electrode Potentials of Zinc and Copper

Discussion

A typical galvanic cell consists of two electrodes connected in series. In theory any reaction involving a transfer of electrons is capable of providing an electrode, i.e. any redox system. In practice, however, to be satisfactory an electrode must be thermodynamically reversible.

The potential of a single electrode cannot be measured experimentally, but when two electrodes form a complete cell the potential of one of them relative to the potential of the other may be measured. In order that electrode potentials may be comparable an arbitrary zero of potential has been established. The potential of a standard hydrogen electrode (hydrogen ions at unit activity and hydrogen gas at unit activity) is taken by convention to be zero at all temperatures and the potential of other electrodes are expressed relative to the potential of the standard hydrogen electrode.

Consider a thermodynamically reversible redox system coupled with a standard hydrogen electrode to form a complete cell:

$$\text{Pt; H}_2\ (a = 1)\ \bigg|\ \text{H}^+\ (a = 1)\ \bigg\|\ \begin{array}{c}\text{oxidized form } (a = a_{\text{ox}}) \\ \text{reduced form } (a = a_{\text{red}})\end{array}\ \bigg|\ \text{Pt}$$

Suppose the cell reaction is

$$m\,\text{ox} + \tfrac{1}{2}\text{H}_2 \to n\,\text{red} + \text{H}^+$$

The free energy increase associated with this reaction will be given by the van't Hoff reaction isotherm

$$\Delta G = \Delta G^0 + RT \ln \frac{a_{\text{red}}^n \times a_{\text{H}^+}}{a_{\text{ox}}^m \times a_{\text{H}_2}^{1/2}} \tag{1}$$

where the superscript 0 refers to the standard state.

As the hydrogen electrode is in the standard state, equation (1) reduces to

$$\Delta G = \Delta G^0 + RT \ln \frac{a_{\text{red}}^n}{a_{\text{ox}}^m} \tag{2}$$

106

If the cell reaction is as stated above the redox electrode will be the positive pole of the cell and the cell e.m.f. E_{cell} will be given by[1]

$$E_{cell} = E_{redox} - E^0_{H_2}$$

but since

$$E^0_{H_2} = 0, \quad \text{then} \quad E_{cell} = E_{redox}$$

where E_{redox} is the potential of the redox system

Now

$$\Delta G = -zFE_{cell} \quad \text{and} \quad \Delta G^0 = -zFE^0_{cell}$$

hence

$$\Delta G = -zFE_{redox} \quad \text{and} \quad \Delta G^0 = -zFE^0_{redox}$$

Substituting in equation (2) we have

$$E_{redox} = E^0_{redox} + \frac{RT}{zF} \ln \frac{a^m_{ox}}{a^n_{red}} \tag{3}$$

Equation (3) is a universal equation relating the potential of an electrode to the activities of the oxidized and reduced species from which it is formed.

In order to obviate the practical difficulties associated with the hydrogen electrode, use is frequently made of a subsidiary reference electrode, the potential of which on the hydrogen scale is accurately known. In this experiment the dip type saturated calomel half cell is employed as a reference electrode. It consists of mercury in contact with an aqueous solution which is saturated with respect both to potassium chloride and mercurous chloride.

If a zinc electrode is combined with a calomel electrode to form a complete cell, the e.m.f. of this cell may be measured and knowing the potential of the calomel electrode the e.m.f. of the zinc electrode may be calculated. The reaction occurring at the zinc electrode is

$$Zn \rightleftharpoons Zn^{2+} + 2e$$

and application of equation (3) to this reaction gives

$$E_{Zn} = E^0_{Zn} + \frac{RT}{2F} \ln m_{Zn^{2+}} \gamma_{\pm} \tag{4}$$

where $\gamma_{\pm}$ is the mean ionic activity coefficient of the zinc sulphate solution and $m_{Zn^{2+}}$ is the molality of the zinc ions. Knowing the potential of the zinc electrode and the activity of the zinc ions in solution the standard electrode potential of zinc may be calculated.

This experiment may be repeated with a copper electrode when the electrode reaction is

$$Cu \rightleftharpoons Cu^{2+} + 2e$$

and equation (3) takes the form

$$E_{Cu} = E_{Cu}^{0} + \frac{RT}{2F} \ln m_{Cu^{2+}} \gamma_{\pm} \tag{5}$$

Apparatus and Chemicals

Muirhead potentiometer, standard cell, galvanometer, accumulator, thermostat, dip-type saturated calomel electrode, copper electrode, zinc electrode and 0·1 M and 0·01 M solutions of copper and zinc sulphates.

Method

Some 0·1 M zinc sulphate solution is added to a beaker and the zinc and calomel electrodes are placed in the solution. The complete cell is then placed in a thermostat at 25°C, allowed to come to thermal equilibrium and connected to the potentiometer. The e.m.f. of the cell is measured and this measurement is repeated after an interval of time to ensure that the system has reached equilibrium.

The process is repeated using 0·01 M zinc sulphate solution.

The experiments are also carried out with a copper electrode instead of a zinc electrode, using firstly 0·1 M copper sulphate solution and secondly 0·01 M copper sulphate solution.

The copper and zinc electrodes must be thoroughly cleaned with acid, washed with water and dried before each experiment.

When using the zinc–calomel cells the potential of the zinc electrode may be deduced from the relationship

$$E_{cell} = E_{cal} - E_{Zn}$$

and for the copper–calomel cells

$$E_{cell} = E_{Cu} - E_{cal}$$

where E_{cal} is given by the empirical relation

$$E_{cal} = 0·244 - 0·0007(t - 25)$$

t being the temperature in °C.

Using the values of the mean ionic activity coefficients given below the standard electrode potentials of zinc and copper are calculated using equations (4) and (5).

The mean activity coefficients ($\gamma_\pm$) at 25°C of the copper and zinc sulphate solutions are

	0·01 M	0·1 M
copper sulphate	0·404	0·158
zinc sulphate	0·400	0·161

[1] A. R. DENARO, *School Sci. Rev.*, **42**, 389; **43**, 135 (1961).

Concentration Cells

Discussion

When two electrodes of the same chemical nature are combined to form a cell, an e.m.f. is produced if there is some difference in activity between the two electrodes. Such a cell is called a concentration cell, and the difference in activity may be brought about by having similar electrodes in contact with solutions of different ionic activities.

As with any other electrochemical cell, the e.m.f. of the cell will be equal to the difference between the potentials of the individual electrodes. This fact may be checked by measuring the potentials of the two electrodes against a reference electrode and subsequently combining the two electrodes to form a concentration cell and measuring the e.m.f. The two electrodes may be conveniently connected by means of a salt bridge which usually consists of a glass tube containing a concentrated solution of a uni-univalent electrolyte in which the transport numbers of the cation and anion are very nearly equal. In this case, the complete cell is known as a concentration cell without transference. In this experiment the following concentration cell is investigated

$$\text{Ag}\Big|\text{AgNO}_3(0\cdot01\ \text{M})\Big|\overset{\text{satd.}}{\text{NH}_4\text{NO}_3}\Big|\text{AgNO}_3(0\cdot1\ \text{M})\Big|\text{Ag}$$

The electrode equilibrium in a $\text{Ag}|\text{AgNO}_3$ half cell is

$$\text{Ag} = \text{Ag}^+ + e$$

and from equation (3) Experiment 44, the potential E^1_{Ag} of the $\text{Ag}|\text{AgNO}_3$ ($0\cdot1$ M) electrode will be given by

$$E^1_{\text{Ag}} = E^0_{\text{Ag}} + \frac{RT}{F}\ln a_{\text{Ag}^+(0\cdot1)} \tag{1}$$

and the potential E^2_{Ag} of the $\text{Ag}|\text{AgNO}_3$ ($0\cdot01$ M) electrode will be given by

$$E^2_{\text{Ag}} = E^0_{\text{Ag}} + \frac{RT}{F}\ln a_{\text{Ag}^+(0\cdot01)} \tag{2}$$

110

The e.m.f. of the concentration cell E will be given by

$$E = E^1_{Ag} - E^2_{Ag} = \frac{RT}{F} \ln\frac{a_{Ag^+(0.1)}}{a_{Ag^+(0.01)}} \tag{3}$$

Apparatus and Chemicals

Muirhead potentiometer, standard cell, galvanometer, accumulator, dip-type saturated calomel electrode, salt bridge, silver electrodes, 0·1 M silver nitrate solution, 0·01 M silver nitrate solution and saturated ammonium nitrate solution.

Method

The silver electrodes are sensitized by dipping them into moderately concentrated nitric acid, containing a little sodium nitrate, until gassing commences. They are then removed and washed well with distilled water. One electrode is then immersed in about 25 ml of 0·1 M silver nitrate solution. The calomel electrode is placed in a beaker containing saturated potassium chloride solution or saturated ammonium nitrate solution. The two electrode solutions are then connected by means of a salt bridge containing saturated ammonium nitrate solution. (The salt bridge is inserted to prevent the precipitation of silver chloride in the calomel electrode.) The e.m.f. of the complete cell is then measured with the potentiometer. The procedure is repeated using 0·01 M silver nitrate solution instead of 0·1 M silver nitrate solution.

The two silver half cells are then connected together using the ammonium nitrate salt bridge and the e.m.f. of the complete cell is measured.

Using the data given below the standard electrode potential of silver E^0_{Ag} is calculated by application of equations (1) and (2).

The e.m.f. of the silver concentration cell is compared with the difference in the potentials of the individual silver electrodes E^1_{Ag} and E^2_{Ag} as measured with the calomel reference electrode. The experimental value of the e.m.f. of the concentration cell is compared with the theoretical value as calculated from equation (3).

$$E_{cal} = 0·244 - 0·0007(t - 25)$$

where t is the temperature in °C.

At 25°C

$$a_{Ag^+(0.1)} = 0·0733$$

$$a_{Ag^+(0.01)} = 0·00892$$

Solubility Product from e.m.f. Measurements

Discussion

The solubility product of an electrolyte may be determined from e.m.f. measurements. A cell is devised in which the potential of one of the electrodes depends upon the activity of one of the ions of the electrolyte under investigation. In the determination of the solubility product of silver chloride, for example, the following concentration cell may be used

$$\text{Ag}\left|\text{AgCl(s)}\;0\cdot01\,\text{M}\,\text{KCl}\right|\underset{\text{NH}_4\text{NO}_3}{\overset{\text{satd.}}{}}\left|0\cdot01\,\text{M}\,\text{AgNO}_3\right|\text{Ag}$$

The potential of the silver–silver chloride electrode will be governed by the activity of the silver ions, a_{Ag^+} in solution which in turn will be governed by the activity of the chloride ions a_{Cl^-}, according to the solubility product principle.

Taking the activity of the silver ions at the silver–silver nitrate electrode as $a_{\text{Ag}^+(0\cdot01)}$ the e.m.f. of the cell E is given by

$$E = \frac{RT}{F}\ln\frac{a_{\text{Ag}^+(0\cdot01)}}{a_{\text{Ag}^+}} \tag{1}$$

The solubility product of silver chloride K_{sp} is defined by

$$K_{\text{sp}} = a_{\text{Ag}^+}\times a_{\text{Cl}^-} \tag{2}$$

hence

$$a_{\text{Ag}^+} = \frac{K_{\text{sp}}}{a_{\text{Cl}^-}}$$

substituting for a_{Ag^+} in equation (1)

$$E = \frac{RT}{F}\ln a_{\text{Ag}^+(0\cdot01)} - \frac{RT}{F}\ln K_{\text{sp}} + \frac{RT}{F}\ln a_{\text{Cl}^-}$$

Converting to common logarithms and rearranging

$$\log K_{\text{sp}} = \log a_{\text{Ag}^+(0\cdot01)} + \log a_{\text{Cl}^-} - \frac{EF}{2\cdot303RT} \tag{3}$$

112

The solubility product may thus be calculated from the e.m.f. of the cell if $a_{Ag^+(0.01)}$ and a_{Cl^-} are known. It is usually assumed that the activity of the chloride ions arises solely from the potassium chloride and that the contribution from the silver chloride is negligible in comparison.

If silver chloride is dissolved in pure water

$$a_{Ag^+} = a_{Cl^-}$$

and hence from equation (2)

$$a_{Ag^+} = a_{Cl^-} = \sqrt{(K_{sp})}$$

If a solution of a sparingly soluble salt is sufficiently dilute for its mean ionic activity coefficient to be assumed to be equal to unity, the solubility of the salt is given by the square root of its solubility product.

Apparatus and Chemicals

Muirhead potentiometer, standard cell, galvanometer, accumulator, salt bridge, silver electrodes, 0·01 M silver nitrate solution, 0·01 M potassium chloride solution and saturated ammonium nitrate solution.

Method

The silver electrodes are sensitized as described in Experiment 45.

One of the electrodes is placed in about 25 ml of 0·01 M potassium chloride solution and two drops of 0·1 M silver nitrate solution are added to give a precipitate of silver chloride. The other silver electrode is placed in about 25 ml of 0·01 M silver nitrate solution. The two electrode solutions are connected with an ammonium nitrate salt bridge and the e.m.f. of the complete cell is measured with the potentiometer.

Given the following data the solubility product of silver chloride is calculated by applying equation (3). From the result the solubility of silver chloride in pure water is also calculated.

At 25°C

$$a_{Ag^+(0.01)} = 0·00892$$

$$a_{Cl^-} = 0·00902$$

The Quinhydrone Reference Electrode

Discussion

In addition to the hydrogen and calomel electrodes a number of other reference electrodes are available. The quinhydrone is perhaps the most readily set up and convenient of all electrodes. It consists of a platinum contact immersed in a buffer solution containing a little quinhydrone which is an equimolecular mixture of hydroquinone and quinone. The electrode equilibrium is

$$Q + 2H^+ + 2e \rightleftharpoons QH_2 \tag{a}$$

and its potential is given by equation (3), Experiment 44.

$$E_{QH_2} = E^0_{QH_2} + \frac{RT}{2F} \ln \frac{a_Q \cdot a^2_{H^+}}{a_{QH_2}}$$

$$= E^0_{QH_2} + \frac{RT}{2F} \ln \frac{a_Q}{a_{QH_2}} + \frac{RT}{F} \ln a_{H^+}$$

or

$$E_{QH_2} = \text{constant} + \frac{2 \cdot 303 RT}{F} \log a_{H^+} \tag{1}$$

assuming that the term $\log a_Q/a_{QH_2}$ is a constant. The constant term in equation (1) is specific for this electrode.

In this experiment a dip-type calomel electrode is connected in series with a quinhydrone electrode system containing a buffer solution of nominal value pH 4. The resultant cell e.m.f. is measured and

$$E_{cell} = E_{QH_2} - E_{cal}$$

Hence from equation (1)

$$pH = \frac{\text{constant} - E_{cal} - E_{cell}}{2 \cdot 303 \, RT/F} \tag{2}$$

Knowing the value of E_{cal}, the quinhydrone electrode constant and RT/F at the temperature of the experiment, the pH of the buffer solution can be calculated. This experiment can be repeated using buffer solutions of nominal pH values in the region of 1 to 8; ionization of the hydroquinone occurs at the higher pH values.

Apparatus and Chemicals

Potentiometer, standard cell, galvanometer, accumulator, quinhydrone electrode, dip-type calomel electrode and buffer solutions of pH 2 to 8.

Method

A small quantity of quinhydrone is added to 50 ml of buffer solution contained in a small beaker. A platinum contact is immersed in the buffer solution and the cell completed by connection with a dip type calomel electrode. The cell e.m.f. is measured and the pH of the buffer solution calculated, using equation (2),[1] the value of E_{cal} at the temperature of the experiment is obtained from Experiment 44. This experiment is then repeated using the four buffer solutions in the pH range 1 to 8.

[1] At 18°C (a) $2 \cdot 303 \ RT/F = 0 \cdot 0577 \ V$; (b) the quinhydrone electrode constant $= 0 \cdot 6942 \ V$. The values of these factors at other temperatures can be obtained from E. C. POTTER, *Electrochemistry, Principles and Applications*, Cleaver-Hume, London (1956).

E

pH Titration Curve

Discussion

One form of the glass electrode consists of a glass tube terminating in a thin walled glass bulb containing a platinum contact immersed in a buffer solution. Its potential (E_G) when immersed in a test solution is dependent on the pH of this same test solution and is given by the equation

$$E_G = \text{constant} + \frac{RT}{F} \ln a_{H^+}$$

where a_{H^+} is the activity of the hydrogen ions in the test solution. The precise potential of this electrode is not of great interest as it includes an asymmetry potential (possibly due to strain at the glass/solution interface) which exists across the glass membrane even when the two solutions on either side of the membrane have identical activity values. In practice, however, one can record the pH throughout a given titration using the cell assembly shown below. Consequently this pH can be used as an "indicator" in acid–alkali titrations, there being a relatively large change in pH at the end point.

In this experiment the glass electrode is connected in series with a saturated dip-type calomel electrode as shown:

solution of constant pH | glass membrane | solution of unknown pH |
| calomel electrode

The change in pH during the titration of sodium hydroxide with hydrochloric acid is recorded. The Poggendorf type bridge cannot be used as the glass electrode resistance is high, i.e. of the order of $10^7 - 10^8 \ \Omega$, and hence the pH is measured using a valve voltmeter (pH meter).

Apparatus and Chemicals

Cambridge bench pH meter, glass electrode, dip-type calomel electrode, 0·1 M hydrochloric acid solution, nominal 0·1 M sodium hydroxide solution.

Method

The electrodes are washed in distilled water, dried with filter paper and washed again by successive immersions in the sodium hydroxide

116

solution. The electrodes are immersed in 25 ml of sodium hydroxide solution contained in a clean beaker. This solution is then titrated with the hydrochloric acid and the pH recorded after each addition of titrant. The burette is arranged so that the jet can be removed from above the solution after each addition. This is to ensure that titrant

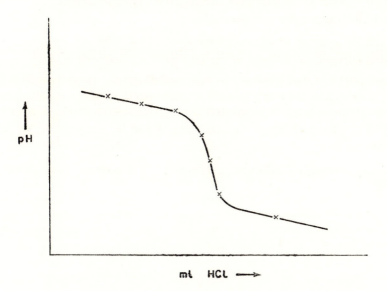

Fig. 1. Variation of pH with added volume of hydrochloric acid.

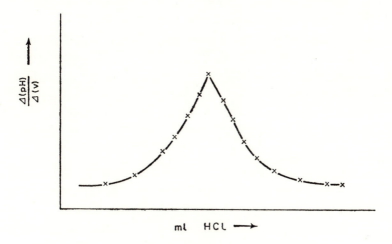

Fig. 2. Variation of the pH differential with added volume of hydrochloric acid.

drops will not fall into the mixture, and it will also permit a more efficient stirring of the mixture before taking a pH reading. A preliminary titration is performed in order to determine the region where the pH changes most rapidly. In subsequent titrations more readings can be taken in this region. On completion of the experiment the electrode system is washed and immersed in distilled water.

The following graphs are then plotted:

(a) pH vs. volume of hydrochloric acid added (Fig. 1).

(b) $\Delta(pH)/\Delta(v)$ vs. volume of hydrochloric acid added (Fig. 2).

The end point of the titration is obtained from the peak as indicated in Fig. 2. The molarity of the sodium hydroxide solution is calculated. A table of indicators[1] is consulted and by reference to their working ranges it is decided which indicator(s) would be preferable for this titration.

[1] *Handbook of Chemistry and Physics* (edited by C. D. HODGMAN), Chemical Rubber Publishing Co.

Study of Potentiometric and Indicator End Points

Discussion

The objects of this experiment are to correlate electrometric and indicator end points, to distinguish between equivalence points and end points and to investigate any "indicator blanks" and hence the choice of indicator.

The titration selected is that of phosphoric acid with sodium hydroxide. Phosphoric acid is a tribasic acid with three successive ionizations:

$$H_3PO_4 = H_2PO_4^- + H^+$$

$$H_2PO_4^- = HPO_4^{2-} + H^+$$

$$HPO_4^{2-} = PO_4^{3-} + H^+$$

In this experiment three titrations of phosphoric acid with sodium hydroxide are carried out using three different indicators. A fourth titration is performed, when the pH is recorded using a pH meter and a suitable glass and calomel electrode assembly.

Apparatus and Chemicals

Cambridge bench model pH meter, glass electrode, dip-type calomel electrode, accumulators, 0·2 M phosphoric acid, 0·5 M sodium hydroxide, methyl red, phenolphthalein and thymolphthalein indicators.

Method

Ten millilitres of the phosphoric acid and 50 ml of distilled water are placed in a 250 ml conical flask and titrated with sodium hydroxide to the methyl red end point. The titration is repeated using phenolphthalein and thymolphthalein. Again 10 ml of the phosphoric acid are placed in a 250 ml beaker and sufficient distilled water added so that the glass and calomel electrodes are just covered. The electrodes are then connected to the pH meter. A preliminary titration is then performed to determine the region where the pH changes most rapidly. In subsequent titrations more frequent observations may then be made in these regions. The solution should be stirred after each addition of alkali before the pH is measured. On completion of the titrations, the electrode system is immersed in distilled water.

The results are recorded in tabular form under the headings:

Volume of sodium hydroxide, v ml	pH	$\dfrac{\Delta(\text{pH})}{\Delta v}$

Two graphs are plotted:

    (a) pH vs. v ml of sodium hydroxide.
    (b) $\Delta(\text{pH})/\Delta v$ vs. v ml. of sodium hydroxide.

The points of inflection on the first graph are correlated with the sharp peaks on the second graph. The volume of sodium hydroxide is determined and the pH observed at the two equivalence points.

A table of indicators[1] is consulted and by reference to their working ranges it is decided which indicator is preferable for the titration of phosphoric acid solution to the first and second equivalence points. Indicator and electrometric end points are compared and indicator blanks, if any, are determined. It should be explained why only two equivalence points are found in the titration of this tribasic acid.

The respective dissociation constants of phosphoric acid are:

$$K_1 = \frac{[\text{H}^+][\text{H}_2\text{PO}_4^-]}{[\text{H}_3\text{PO}_4]} = 7\cdot54 \times 10^{-3}$$

$$K_2 = \frac{[\text{H}^+][\text{HPO}_4^{2-}]}{[\text{H}_2\text{PO}_4^-]} = 6\cdot23 \times 10^{-8}$$

$$K_3 = \frac{[\text{H}^+][\text{PO}_4^{3-}]}{[\text{HPO}_4^{2-}]} = 4\cdot79 \times 10^{-13}$$

---

[1] C. D. HODGMAN (Ed.), *Handbook of Chemistry and Physics*, Chemical Rubber Publishing Co.

Potentiometric Titration—Verification of the Nernst Equation

Discussion

When using a suitable cell oxidation–reduction titrations may be carried out potentiometrically, for on recording the variation of the e.m.f. of the cell with volume of reagent added, a sharp change in e.m.f. occurs at the equivalence point. In this experiment ferrous iron is oxidized to ferric iron by the addition of potassium dichromate solution. The end point of the titration is determined in two solutions of different acid concentration.

When a platinum electrode is immersed in an oxidation reduction system (e.g. ferrous–ferric) a potential is established. In the case of the ferrous–ferric system, the electrode reaction is

$$Fe^{3+} + e \rightleftharpoons Fe^{2+}$$

and its potential is given by equation (3), Experiment 44.

$$E_{Fe^{3+}/Fe^{2+}} = E^0_{Fe^{3+}/Fe^{2+}} + \frac{RT}{F} \ln \frac{a_{Fe^{3+}}}{a_{Fe^{2+}}} \tag{1}$$

A cell consisting of a platinum and a dip-type calomel electrode immersed in a ferrous ammonium sulphate solution can be used and the e.m.f. of the cell (E_{cell}) is recorded as the ferrous solution is titrated with potassium dichromate solution. If the ferrous solution and the dichromate solution are of the same normality, the ratio $[Fe^{3+}]/[Fe^{2+}]$ is given by

$$\frac{[Fe^{3+}]}{[Fe^{2+}]} \triangleq \frac{x}{a-x} \tag{2}$$

where x is the volume of added dichromate at any instant and a is the initial volume of ferrous ammonium sulphate solution.

Substituting this ferrous–ferric ratio in equation (1),

$$E_{Fe^{3+}/Fe^{2+}} = E^0_{Fe^{3+}/Fe^{2+}} + \frac{RT}{F} \ln \frac{x}{a-x} \tag{3}$$

The value of $E_{Fe^{3+}/Fe^{2+}}$ may thus be calculated. If equation (1) is valid a plot of $E_{Fe^{3+}/Fe^{2+}}$ against E_{cell} should be linear as the potential of the calomel electrode is constant. It should be noted that in the calculation concentration values of the ferrous and ferric ions are used instead of activities.

Apparatus and Chemicals

Cambridge pH meter (millivoltmeter), calomel electrode, platinum electrode, 0·1 N potassium dichromate solution, 0·1 N ferrous ammonium sulphate solution and "bench" dilute sulphuric acid.

Method

The electrodes are thoroughly washed in distilled water, dried and then immersed in a 400 ml beaker containing 25 ml of the ferrous ammonium sulphate solution, 50 ml of dilute sulphuric acid and 50 ml of distilled water. A preliminary titration is performed with the potassium dichromate solution, and the region in which the e.m.f. of the cell changes most rapidly is determined. In subsequent titrations, more readings are made in this region. General titration instructions are given in Experiment 48.

On completion of this titration, the electrode assembly is washed and the titration repeated using 25 ml of ferrous ammonium sulphate solution, and 100 ml of dilute sulphuric acid.

The results are recorded in tabular form under the headings:

Volume potassium dichromate added x	Volume ferrous soln. remaining $a-x$	$\dfrac{x}{a-x}$	E_{cell} (measured)	$E_{calc.}$ (from equation 3)

The following graphs are plotted:

(a) E_{cell} vs. volume of potassium dichromate added;

(b) $\Delta E/\Delta(v)$ vs. volume of potassium dichromate added;

(c) E_{cell} vs. $E_{calc.}$.

From the graphs (a) and (b) the equivalence points for the titrations are determined, see Experiment 48. The effect of increasing the acidity of the solution in the second titration should be discussed. Similarly the linear property of graph (c) should be discussed from the viewpoint of the validity of the Nernst equation.

At 17°C

$$2\cdot303\frac{RT}{F} = 0\cdot0575 \text{ V}$$

$$E^0_{Fe^{3+}/Fe^{2+}} = 0\cdot712 \text{ V}$$

Thermodynamics of Cells

Discussion

The e.m.f. of a voltaic cell depends upon temperature. From the temperature coefficient of e.m.f. and a knowledge of the e.m.f. at a particular temperature, three thermodynamic quantities associated with the reversible reaction occurring in the cell can be evaluated. These thermodynamic quantities are the free energy increase ΔG, the entropy increase ΔS, and the enthalpy increase ΔH given by equations (1), (2) and (3) severally.

$$\Delta G = -zFE \tag{1}$$

$$\Delta S = zF\left(\frac{\partial E}{\partial T}\right)_p \tag{2}$$

$$\Delta H = zF\left[T\left(\frac{\partial E}{\partial T}\right)_p - E\right] \tag{3}$$

where E is the e.m.f. of the cell, z the number of electrons involved in the reaction, T the absolute temperature and $(\partial E/\partial T)_p$ the temperature coefficient in volts per degree Centigrade.

A suitable cell for study is

$$\text{Zn} - \text{Hg} \mid \text{Zn Cl}_2 \ (0\cdot 5m), \text{AgCl} \ (s) \mid \text{Ag}$$

which may be assembled as shown in Fig. 1.

The cell reaction is

$$\text{Zn} + 2\text{AgCl}(s) \rightarrow 2\text{Ag} + \text{Zn}^{2+} + 2\text{Cl}^-$$

Apparatus and Chemicals

Tinsley potentiometer, standard cell, accumulator, galvanometer, water bath, amalgamated zinc electrode, silver–silver chloride electrode, zinc chloride, 250 ml wide mouthed conical flask, thermometer.

Method

The zinc electrode is prepared by rubbing a zinc rod with a drop of mercury using cotton wool under dilute nitric acid in a small beaker. The electrode is transferred to decinormal hydrochloric acid and

E*

123

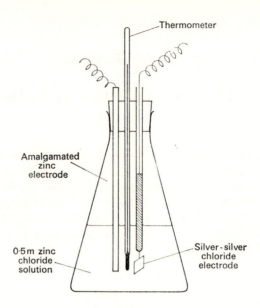

Fɪɢ. 1. The cell.

rubbed again until the surface is uniformly bright. The electrode is kept under dilute hydrochloric acid until required.

The silver–silver chloride electrode is best prepared as follows. A platinum electrode is cleaned with nitric acid. It is then used as the cathode in a silver plating solution at a current density of about 5 mA cm^{-2} for about 10 to 15 min. (The silver plating solution is prepared by dissolving 10 g of potassium argentocyanide in 1 l. of water and adding silver nitrate solution until a faint cloudiness appears.) The result should be a uniformly white deposit of silver over the electrode. The electrode is now washed with distilled water and used as an anode in the electrolysis of 1 M hydrochloric acid with a platinum cathode. With a current density of about 5 mA cm^{-2} for about 10 to 15 min the electrode should have a uniform chocolate-coloured coating of silver chloride. It should be kept under dilute hydrochloric acid until required.

0·5 *m* zinc chloride solution is made up by weighing solid sticks of zinc chloride and dissolving them in water. Any precipitate of zinc hydroxide is filtered off. About 150 ml of the zinc chloride solution is placed in the wide-mouthed conical flask and the two electrodes together with a thermometer are fitted into the bung. The cell is placed in a water bath at 20°C and connected to the potentiometer. The flask may be swirled gently to achieve thermal equilibrium or the solution may be stirred magnetically. The e.m.f. of the cell is then measured at frequent intervals, e.g. every 5 min until a constant reading is obtained. The cell is then assumed to be at the same temperature as the thermostat. These measurements are repeated at 25°, 30°, 35° and 40°C.

The procedure is then reversed by taking e.m.f. readings at appropriate temperature intervals as the thermostat cools.

A graph of e.m.f. against t °C is plotted, using both series of results. The value of $(\partial E/\partial T)_p$ at 25°C is determined from the graph and used in conjunction with the mean value of E at this temperature to calculate ΔG, ΔH and ΔS. Comments should be made on the positive or negative character of these thermodynamic quantities. It should be noted that all solubility temperature relationships have been ignored in these calculations.

PART II

The Radius of a Molecule from Viscosity Measurements

Discussion

The ratio of the viscosity η of a solution of large spherical particles to the viscosity η_0 of the pure solvent is related to the fraction ϕ of the total volume of spheres per millilitre of solution by the expression, derived by Einstein,

$$\frac{\eta}{\eta_0} = 1 + 2 \cdot 5\phi$$

This equation may be rewritten in the form

$$\frac{\eta}{\eta_0} = 1 + 6 \cdot 3 \times 10^{21} \, r^3 c$$

where r is the radius of the particle and c the concentration of the particles in moles/l. Hence if η/η_0 is plotted against c a straight line is obtained and r may be determined from the slope.

The relative viscosity η/η_0 may be determined as in Experiment 5 using the relationship

$$\frac{\eta}{\eta_0} = \frac{td}{t_0 d_0}$$

In this equation t and t_0 refer to the time of flow through the capillary of a given volume of solution and pure solvent, respectively, and d and d_0 are the densities of the solution and pure solvent.

Apparatus and Chemicals

Thermostat at 25°C, Cannon–Fenske viscometer, stop watch and glycerol.

Method

A solution is prepared containing 1 g mole of glycerol per l. of solution. The time of flow of a given volume of this solution is determined using the viscometer as in Experiment 4. The time of flow of the same volume of water is then determined.

The glycerol solution is diluted by factors of 0·75, 0·5 and 0·25, and the time of flow measured for each dilution using the same volume as before.

The results are recorded in tabular form under the headings:

c g moles per l.	t sec	$\dfrac{t}{t_0}$	$\dfrac{d}{d_0}$	$\dfrac{\eta}{\eta_0}$

A graph is plotted of relative viscosity vs. concentration and the radius of the glycerol molecule determined.

For glycerol solutions it may be assumed that d/d_0 is equal to $1 + 0 \cdot 021c$.

Runs should be repeated several times until constant values are obtained. This ensures that thermal equilibrium exists. The experiment is repeated with 10 per cent solutions of (a) sucrose–glucose and (b) ethyl alcohol–tert-amyl alcohol.

In the light of the results, the application of Einstein's equation to molecular solutions should be discussed.

Dipole Moment of a Polar Molecule

Discussion

The capacitance of a condenser is defined as the quantity of electricity stored per unit difference of potential between the plates. The unit of capacitance, the farad F, is one coulomb per volt which is equal to 9×10^{11} e.s.u./V. This is rather a large unit and more practical units are the microfarad ($1\ \mu F = 10^{-6}$ F) and the picrofarad ($1\ \mu\mu F = 10^{-12}$ F).

The capacitance depends upon the geometry of the condenser and the nature of the medium between the plates. The dielectric constant ϵ (specific inductive capacity or relative permittivity) is defined by

$$\epsilon = C/C_0$$

where C is the capacitance of the condenser containing the medium, and C_0 the capacitance of the same condenser completely evacuated. Since the dielectric constant is a ratio between similar quantities, it is dimensionless.

For a given substance the dielectric constant is related to total polar polarization P by the equation:

$$P = \frac{\epsilon-1}{\epsilon+2} \cdot \frac{M}{d} \tag{1}$$

where M is the molecular weight and d the density. This property has, therefore, the dimensions of molar volume. It is the sum of the orientation polarization P_{Or} and the distortion polarization P_D

$$P = P_{Or} + P_D \tag{2}$$

The orientation polarization is a function of the permanent dipole moment of the molecule and of the temperature.

$$P_{Or} = \frac{4\pi N}{3} \cdot \frac{\mu^2}{3kT}$$

where μ is the dipole moment of the molecule, T the temperature in degrees Kelvin, N Avogadro's number, and k the Boltzmann constant, (the factor $4/3\pi N$ is required to convert the molecular quantity

$(\mu^2/3kT)$ into the required molar quantity). Hence the dipole moment is given by

$$\mu = \left[\frac{9kT}{4\pi N} \cdot P_{\mathrm{or}}\right]^{1/2} \tag{3}$$

When the Boltzmann constant k is expressed in ergs per degree per molecule, the value of the dipole moment is in electrostatic-cm units. As the moment is the product of an electrostatic charge and a distance and as the former is of the order of 10^{-10} e.s.u. and the latter 10^{-8} cm, a convenient unit should be equal to 10^{-18} electrostatic-cm units. This unit is called a Debye (D).

Hence, in order to calculate the dipole moment, it is necessary to know the orientation polarization P_{Or} at a particular temperature. This quantity may be calculated from equation (2), provided the total molar polarization P, and the distortion polarization P_{D} are known.

The distortion polarization is due to the induced shifts of the atoms or electrons within the molecule and it is related to the refractive index:

$$P_{\mathrm{D}} = \frac{n^2 - 1}{n^2 + 2} \cdot \frac{M}{d} \tag{4}$$

The refractive index of a liquid can be determined by an Abbé refractometer, and the density with a pyknometer. The distortion polarization may then be calculated from these measurements. Alternatively, as it is an additive and constitutive property, it may be calculated from known values of atomic and structural refractions.

The total molar polarization P is more difficult to determine. For a non-polar substance it may be calculated from a knowledge of the dielectric constant (ϵ), the molecular weight (M), and the density (d), equation (1). For a polar substance, the polar molecules tend to orientate so that the opposite charged ends are adjacent, and incorrect results are obtained. This dipole–dipole interaction is minimized if measurements are made in various dilute solutions and the results extrapolated to infinite dilution.

If $P_{1,2}$ is the total molarization of the solution, x the mole fraction and 1 and 2 refer to the solvent and solute, respectively, then

$$P_{1,2} = x_1 P_1 + x_2 P_2$$
$$= (1 - x_2) P_1 + x_2 P_2$$
$$P_2 = P_1 + (P_{1,2} - P_1)/x_2$$

In order to calculate P_2 at infinite dilution P_2^0 a number of methods have been suggested. According to a method suggested by Hedestrand,[1] if the dielectric constant and the density of a series of solutions of the solute vary linearly with the molar fractions of the solute, then

$$\epsilon_{1,2} = \epsilon_1(1 + ax_2) \tag{5}$$

$$d_{1,2} = d_1(1 + bx_2) \tag{6}$$

The coefficients a and b can be determined from plots of $\epsilon_{1,2}$ vs. x_2 and $d_{1,2}$ vs. x_2. The molar polarization at infinite dilution can then be determined from the relationship:

$$P_2^0 = \frac{3a\epsilon_1}{(\epsilon_1 + 2)^2} \cdot \frac{M_1}{d_1} + \frac{\epsilon_1 - 1}{\epsilon_1 + 2} \frac{1}{d_1}(M_2 - bM_1) \tag{7}$$

Once the total polarization of the solute at infinite dilution P_2^0 is known, the orientation polarization can be calculated

$$P_{\text{or}} = (P_2^0 - P_D) \tag{8}$$

Apparatus and Chemicals

Wayne Kerr Universal bridge,[2] Marconi dielectric constant cell, refractometer, pyknometer with caps, pure benzene, pure mercury, pure polar compound.

Method

For the Marconi dielectric cell the relationship between the measured capacitance C_m, between the terminals and the required capacitance C_r, of the concentric cylindrical cell is

$$C_m = C_r + C_s \tag{9}$$

where C_s is the additional stray capacitance of the terminals. This stray capacitance is a constant for the cell.

The capacitance C_m^0 is measured with air in the cell. The capacitance C_m^b is measured with 12 ml of anhydrous benzene in the cell. Then as

$$C_m^b = 2 \cdot 284 C_r + C_s \tag{10}$$

and

$$C_m^0 = 1 \cdot 000 C_r + C_s \tag{11}$$

$$(C_m^b - C_m^0) = 1 \cdot 284 C_r \tag{12}$$

Hence the value of C_r can be calculated. Substitution of this value into equation (10) or (11) gives the value of C_s.

The value of the required capacitance of the air C_r^0 and the required capacitance of the benzene C_r^b can be found from

$$C_r^0 = C_m^0 - C_s$$

and

$$C_r^b = C_m^b - C_s$$

Hence the dielectric constant ϵ_1 for the benzene can be calculated

$$\epsilon_1 = C_r^b / C_r^0$$

The capacitance of a 1 mole per cent solution of the polar substance in benzene can be found by adding 12 ml of the solution to the cell and measuring the capacitance C_m^s. Then as before we have

$$C_r^s = C_m^s - C_s$$

and the dielectric constant of the solution $\epsilon_{1,2}$ can be calculated

$$\epsilon_{1,2} = C_r^s / C_r^o$$

The measurement is repeated with 12 ml samples of 2, 3 and 4 mole per cent solutions of the polar substance in benzene. The densities of benzene d_1, and of each of the solutions of the polar substance in benzene $d_{1,2}$ are determined with a pyknometer.

Graphs of $\epsilon_{1,2}$ vs. x_2, and $d_{1,2}$ vs. x_2 are drawn and from these graphs the coefficients a and b can be determined. The molar polarization P_2^0 at infinite dilution can be calculated from equation (7).

The refractive index of the polar liquid is determined with a refractometer and the distortion polarization calculated from equation (4).

The orientation polarization is calculated from equation (8) and the dipole moment from equation (3).

[1] G. HEDESTRAND, *Z. phys. Chem.*, **2**, 428 (1929).

[2] K. A. FLETCHER, *Automatic Measurements of Quality in Process Plants*, The measurement of dielectric constant applied to the quality control of materials, Butterworth Scientific Publications. Details for the use of the bridge are given in the manufacturer's manual.

The Additivity of Group Moments in Aromatic Compounds

Discussion

To a certain extent the dipole moments of groups of atoms may be added as vectors in molecules. In order to test the additivity of group moments it is necessary to attach two or more polar groups to a rigid structure so that certain geometrical relationships between them will be maintained. Benzene is a symmetrical planar molecule with the bonds to the hydrogen atoms disposed radially from the centre of the ring in the plane of the ring. As benzene is a regular hexagon the angle between a pair of *ortho*-substituents is 60°, a pair of *meta*-substituents is 120°, and a pair of *para*-substituents 180°. Benzene itself has zero dipole moment but if one of the hydrogen atoms is replaced by an atom or group of atoms (other than phenyl) the compound will possess a dipole moment. The magnitude of this dipole moment should be the vector sum of the moments of the bond introduced and that of the C—H bond in the *para*-position. If the latter is taken, arbitrarily, as zero, and the dipole moments of a number of benzene substituents are known, it is possible to calculate group moments for aromatic compounds.

The moments produced when chloro and nitro groups are introduced into the benzene ring are, respectively, 1·58 D and 3·01 D units. These values are characteristic of the group and are known as group moments. Using these values, the dipole moments of the disubstituted benzene derivatives:

| ortho | meta | para |

can be calculated from the equations

$$\mu_{ortho} = \sqrt{(\mu_1^2 + \mu_2^2 + 2\mu_1\mu_2 \cos 60°)} \qquad (1)$$

$$\mu_{meta} = \sqrt{(\mu_1^2 + \mu_2^2 + 2\mu_1\mu_2 \cos 120°)} \tag{2}$$

$$\mu_{para} = \mu_1 - \mu_2 \tag{3}$$

The validity of this relationship for vector addition may be tested with *ortho*-and *para*-nitrochlorobenzene. Additional disubstituted derivatives may be studied if time and materials are available.

Apparatus and Chemicals

Heterodyne beat oscillator, pyknometer with caps, refractometer, pure benzene, pure *ortho*- and *para*-nitrochlorobenzene.

Method

For the accurate measurement of the dipole moment of a liquid, probably the best available method is the heterodyne beat method. Exact details for the circuit, etc., are given by Chein[1] and Smyth.[2]

The dipole moments of the disubstituted derivatives of benzene are determined using the heterodyne beat method by the procedure described in Experiment 53. The experimental value for the dipole moment is compared with that calculated by vector addition of the group moments using equations (1) and (3).

[1] J. Y. CHEIN, *J. Chem. Educ.*, **24**, 494 (1947).
[2] SMYTH, in *Technique of Organic Chemistry* (edited by WEISSBERGER), 2nd ed., vol. 1, pt. II, ch. 24, Interscience Publishers (1949).

The Joule–Thomson Coefficient

Discussion

When a gas expands through a throttled tube under adiabatic conditions, the enthalpy of the system remains constant. At the same time the gas cools and the rate of change of temperature with pressure at constant enthalpy is expressed as $(\partial T/\partial p)_H$. This quantity is known as the Joule–Thomson coefficient and it can be shown that the following relationship is valid for all fluids

$$\mu = -\frac{1}{C_p}\left[\left(\frac{\partial U}{\partial v}\right)_T\left(\frac{\partial v}{\partial p}\right)_T + \left(\frac{\partial(pv)}{\partial p}\right)_T\right] \tag{1}$$

where μ is the Joule–Thomson coefficient, C_p the heat capacity of the fluid at constant pressure and U the internal energy of the fluid. For a perfect gas $(\partial U/\partial v)_T = 0$, and $(\partial(pv)/\partial p)_T = 0$, so that $\mu = 0$. The Joule–Thomson coefficient may thus be regarded as a measure of deviation from perfection.

If a gas obeys van der Waals' equation of state the following relationship between the Joule–Thomson coefficient and the constants a and b of van der Waals' equation is approximately true:

$$\mu = \frac{1}{C_p}\left(\frac{2a}{RT} - b - \frac{3abp}{R^2T^2}\right)$$

or at low pressures

$$\mu = \frac{1}{C_p}\left(\frac{2a}{RT} - b\right) \tag{2}$$

Further, it can be shown that

$$\left(\frac{\partial U}{\partial v}\right)_T = \frac{a}{v^2} \tag{3}$$

for a van der Waals gas where v is the gram molecular volume of the gas and hence a/v^2 is the internal pressure term in the van der Waals equation.

If the Joule–Thomson coefficient is determined, from a knowledge of C_p, a and b, equation (2) may be verified. Further, knowing the values of the quantities $(\partial(pv)/\partial p)_T$ and $(\partial v/\partial p)_T$, $(\partial U/\partial v)_T$ may be calculated from equation (1) and equation (3) may be verified.

In this experiment nitrogen is passed along a tube containing a sintered disc and the difference in the temperature of the gas before and after expansion is measured by means of two matched thermistors. The apparatus is illustrated in Fig. 1.

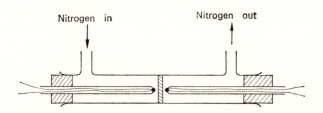

Nitrogen in   Nitrogen out

FIG. 1. Apparatus for the determination of the Joule–Thomson coefficient.

The two thermistors form two arms of a Wheatstone bridge, the remaining two arms being a fixed resistance and a variable resistance. The circuit is shown in Fig. 2, where the thermistors in the inlet and outlet sides of the apparatus are represented as Th_1 and Th_2, respectively. R_f is the fixed resistance and R_v the variable resistance.

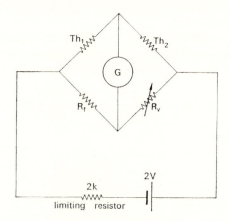

FIG. 2. The Wheatstone bridge circuit.

The relationship between the resistance R of a thermistor and the absolute temperature T is given by

$$R = Ae^{b/T}$$

where b is a constant depending on the material of the thermistor and

A is a constant depending upon its geometry. If two matched thermistors are used they will have been manufactured in the same batch and the value of b for each should be the same. The values of A will probably differ slightly.

The bridge may first be balanced with no gas flowing and consequently no difference in the temperatures of the thermistors. Suppose that under these conditions the resistances of the inlet and outlet thermistors are Th_1 and Th_2, respectively, their temperature being T_0. Suppose also that the value of the variable resistance required for balance is R_0. Then

$$\frac{R_f}{R_0} = \frac{Th_1}{Th_2} = \frac{A_1 e^{b/T_0}}{A_2 e^{b/T_0}} = \frac{A_1}{A_2} \tag{4}$$

where A_1 and A_2 are the values of the constants A for the inlet and outlet thermistors respectively.

If nitrogen is now allowed to flow through the system the temperatures and resistances of the thermistors will change. Suppose that T_L, T_R, Th_1' and Th_2' are the temperatures and resistances of the inlet and outlet thermistors, respectively, under the new conditions. Suppose also that the new value of the variable resistance required to balance the bridge again is R_g. Then

$$\frac{R_f}{R_g} = \frac{Th_1'}{Th_2'} = \frac{A_1 e^{b/T_L}}{A_2 e^{b/T_R}}$$

or

$$\frac{R_f}{R_g} = \frac{A_1}{A_2} e^{b(1/T_L - 1/T_R)}$$

Substituting from equation (4)

$$\frac{R_f}{R_g} = \frac{R_f}{R_0} e^{b(1/T_L - 1/T_R)}$$

or

$$R_g = R_0 e^{-b(1/T_L - 1/T_R)}$$

whence

$$R_g = R_0 e^{b[(T_L - T_R)/T_L T_R]}$$

Putting $T_L - T_R = \Delta T$, which will be small, thus allowing the approximation $T_L T_R \simeq T_0^2$ to be made,

$$R_g = R_0 e^{(b\Delta T/T_0^2)}$$

Once again as ΔT is small the exponential term may be expanded, neglecting terms containing ΔT to the power of two and above.

$$R_g = R_0\left(1 + \frac{b\Delta T}{T_0^2}\right)$$

Let

$$R_g - R_0 = \Delta R$$

then

$$\Delta R = R_0\left(1 + \frac{b\Delta T}{T_0^2}\right) - R_0$$

or

$$\Delta R = R_0 \cdot \frac{b\Delta T}{T_0^2} \tag{5}$$

Apparatus and Chemicals

Tube containing a 10 mm sintered disc, two matched STC Type F23 thermistors (resistance at room temperatures about $2\,k$), manometer, cylinder of nitrogen, fixed resistance, variable resistance, galvanometer, accumulator.

Method

The bridge is first balanced with no nitrogen flowing and the value R_0, of the variable resistor is noted.

The nitrogen is then led through a long copper tube immersed in a thermostat at 25°C and introduced into the glass tube containing the sintered disc. The nitrogen expands through the disc and escapes into the atmosphere. The thermistors are located one on either side of the disc so that the difference in temperature of the gas before and after it has passed the sinter may be deduced. The tube carrying the sinter may be embedded in expanded polystyrene or some other thermal insulator and mounted on a board.

Nitrogen is passed through the apparatus at various initial pressures between 1 and 2 atmospheres as measured by the manometer in the feed line. The various values R_g of the variable resistance required to balance the bridge for each initial pressure of nitrogen are noted.

Assuming that the final pressure is atmospheric, the pressure difference ΔP across the sinter is plotted against ΔR. As the Joule–Thomson coefficient hardly varies over a pressure range of one atmosphere the graph should be a straight line of slope $\Delta P/\Delta R$. Substituting for ΔR from equation (5)

$$\frac{\Delta P}{\Delta R} = \frac{\Delta P}{\Delta T} \cdot \frac{T_0^2}{R_0 b}$$

or

$$\frac{\Delta P}{\Delta T} = \frac{\Delta P}{\Delta R} \cdot \frac{R_0 b}{T_0^2}$$

Using this relationship the slope of the graph is converted into terms of $\Delta P/\Delta T$ which is the reciprocal of the Joule–Thomson coefficient.

Given the following data for nitrogen, equations (2) and (3) should be verified.

$a = 1 \cdot 31$ atm l^2 mole^{-2} $b = 0 \cdot 0373$ l. mole^{-1}

$C_p = 0 \cdot 287$ l. atm °C^{-1} mole^{-1}

$(\partial(pv)/\partial p)_T = 0 \cdot 007$ l. $(\partial v/\partial p)_T = RT$

Molar volume $= 22 \cdot 4$ l. mole^{-1}

The Partition Coefficient in Gas Chromatography

Discussion

In gas–liquid chromatography a substance is distributed between two phases but the process is a continuous one as opposed to a batch operation. One of the phases is a liquid phase which is supported on an adsorbent solid and the other phase is a gas. The liquid phase is stationary while the gas phase moves over it continuously.

The sample is introduced into the gas phase and is caused to vaporize by choosing a suitable temperature for the experiment. As the carrier gas moves over the liquid phase the sample is distributed between the liquid and the gas. As more of the carrier gas, containing none of the sample, passes over the stationary phase which has already dissolved some of the sample, the sample is redistributed between the two phases. In this way the sample is carried along the column of stationary phase until after some delay it is finally eluted and makes its exit from the column once more contained in the mobile phase. By the use of some suitable detector the presence of the sample in the exit gas may be recorded on a moving chart. The usual type of detector (a differential detector) produces a trace as shown in Fig. 1.

O represents the time of injection of the sample and the small peak *A* is due to the presence of air which usually enters the column at the time of injection of the sample. The large peak *P* is due to the sample and the area under the peak is proportional to the quantity of sample injected.

The partition coefficient *K* of the sample between the liquid phase and the gas phase is given by

$$K = \frac{\text{weight of solute per ml of stationary phase}}{\text{weight of solute per ml of mobile phase}}$$

This partition coefficient may be determined from the data on the recorder chart together with certain supplementary data.

It will be convenient here to define certain parameters relating to gas–liquid chromatography.

V_R is the uncorrected retention volume which is the volume of gas required to elute the sample. It is given by

$$V_R = t_R F$$

where F is the flow rate of the carrier gas and t_R is the time required to elute the sample. t_R corresponds to the distance OZ in Fig. 1.

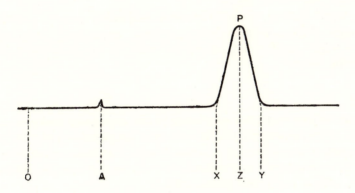

FIG. 1. Recorder chart.

V_M is the gas hold-up which is the uncorrected retention volume of a non-adsorbed sample. It is the volume of carrier gas required to transport such a sample from the point of injection to the detector. Air is such a non-adsorbed sample and hence V_M corresponds to OA.

V_R' is the adjusted retention volume, defined by

$$V_R' = V_R - V_M = AZ \cdot F$$

V_N is the net retention volume, defined by

$$V_N = j V_R' = jAZ \cdot F \tag{1}$$

where

$$j = \frac{3}{2} \left[\frac{(p_i/p_0)^2 - 1}{(p_i/p_0)^3 - 1} \right] \tag{2}$$

j is called the pressure gradient correction factor and p_i and p_0 are the inlet and outlet pressures of the carrier gas, respectively.

The partition coefficient K is related to the net retention volume by

$$K = \frac{V_N d}{w} \tag{3}$$

where d is the density of the stationary phase and w is the weight of stationary phase in the column.

The performance of the column can be expressed in terms of a theoretical plate number n given by

$$n = 16 \frac{(OZ)^2}{(XY)^2} \qquad (4)$$

Apparatus and Chemicals

Gas–liquid chromatography column containing dinonyl phthalate as stationary phase, normal butyl alcohol, secondary butyl alcohol, tertiary butyl alcohol, an equimolar mixture of these three alcohols and a stop clock.

Method

The procedure will vary according to the particular make of chromatograph being employed and it will only be possible here to indicate the necessary operations in a general way. Reference should be made to the manufacturer's instruction manual for the specific methods to be used with a particular instrument.

The inlet pressure of the carrier gas must be measured and this can be done by inserting a manometer in the feed line. The outlet pressure of the gas will usually be atmospheric but if a soap film flowmeter is used in the exit line allowance will have to be made for the vapour pressure of the soap solution. From these data j is calculated using equation (2).

The rate of flow of the carrier gas must also be determined. The flow rates necessary will usually be in the range 20–40 ml/min.

The weight of stationary phase in the column should be known (about 7·0 g). This may be accomplished by adsorbing a known weight of the liquid phase on a known weight of the solid support and then weighing the amount of this mixture required to fill the column.

A sample of alcohol is injected into the column and the net retention volume is calculated using equation (1). Knowing the density of the liquid phase at the temperature of the experiment the partition coefficient of acetone between the mobile phase and the stationary phase may be calculated using equation (3). The theoretical plate number of the column for acetone is calculated directly from the data on the chart using equation (4).

The experiment is repeated using each alcohol successively. The values of the theoretical plate number of the column for each substance are compared. In the mixture the amounts of the three alcohols are equal. The resulting chromatogram should be examined to see whether there is any parameter which has a similar value for all three alcohols and which could therefore be used to estimate the relative quantities present in a mixture.

A reasonable temperature of operation is 50°C at which temperature the density of dinonyl phthalate is given by:

$$d_4^{50} = 0 \cdot 9473 \text{ g ml}^{-1}$$

The size of sample for injection may vary from $0 \cdot 025 \, \mu l$ to $0 \cdot 01$ ml, depending on the sensitivity of the detector.

The Raoult Law Factor in Gas Chromatography

Discussion

The principles of gas–liquid chromatography have been discussed in Experiment 56 and it will be assumed that the student is familiar with them.

The determination of the activity coefficients of solutes in high boiling solvents may be conveniently carried out by gas–liquid chromatography. The information derived from such studies finds its greatest application in the field of gas–liquid chromatography itself and the results may be expressed in two ways. The apparent activity coefficient at infinite dilution or the Raoult law factor γ is given by

$$\gamma = \frac{RTw}{Mp^\circ V_N} \tag{1}$$

where T is the absolute temperature of the column,

w is the weight of stationary phase in the column,

M is the molecular weight of the stationary phase,

p° is the equilibrium vapour pressure of the pure solute in mm of mercury at the column temperature,

V_N is the net retention volume of the solute.

To obtain the true activity coefficient at infinite dilution γ_∞ the Raoult law factor must be adjusted by a term which accounts for the non-ideality of the gas phase. The true and apparent activity coefficients may be regarded as rational and practical coefficients. For many purposes in gas–liquid chromatography γ provides sufficient information but for a thermodynamic analysis of the systems concerned γ_∞ has more meaning.

There is a further retention parameter of great use in gas–liquid chromatography, the specific retention volume V_g which is the net retention volume at 0°C per gram of liquid phase and is given by

$$V_g = \frac{273 V_N}{wT} \tag{2}$$

where the symbols have the same significance as in Experiment 56.

V_g is a constant for a particular substance in a given partitioning system at a fixed temperature. If sufficient data are available the variation of V_g with temperature may be expressed by means of an Antoine equation, viz.

$$\log V_g = A + \frac{B}{T+C}$$

where A, B and C are constants. This parameter may thus be used as a means of identifying samples which are partitioned in a given system. Further, it may be seen that if two substances have appreciably different specific retention volumes it should be possible to separate them quite easily by gas–liquid chromatography.

Apparatus and Chemicals

Gas–liquid chromatography column containing dinonyl phthalate as stationary phase, stop clock, benzene and cyclohexane.

Method

Reference should be made to Experiment 56 for general remarks on the method of procedure. The experiment should be carried out at 50°C.

The flow rate of the carrier gas and its inlet and outlet pressures are determined. A sample of benzene is injected into the column and its net retention volume is calculated. Hence γ for a solution of benzene in dinonyl phthalate is calculated using equation (1). The specific retention volume of benzene under the conditions of the experiment is also calculated from equation (2).

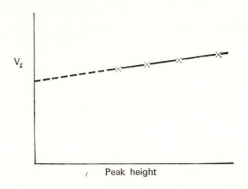

FIG. 1. Variation of specific retention volume with peak height.

F

V_g, the specific retention volume, is often found to vary with sample size, usually decreasing as the sample size increases. It is wise to check this, and if a variation in V_g is found, then V_g should be plotted against the sample size, or, more conveniently, the peak height and the graph extrapolated to zero sample size as in Fig. 1. The value of V_g obtained from this graph is used to calculate the apparent activity coefficient.

These measurements and calculations are repeated with cyclohexane and from a comparison of the specific retention volumes of the two substances examined comment is made on the suitability of gas–liquid chromatography for the separation of a mixture of benzene and cyclohexane. How would these two liquids be separated by other physical or chemical means?

Activity of a Non-electrolyte by Cryoscopy

Discussion

If a solution is sufficiently dilute that the heat of dilution may be neglected, the activity of the solvent a_1 at the freezing point T of the solution is given by

$$\left[\frac{\partial \ln a_1}{\partial T}\right]_p = \frac{\Delta H}{RT^2} \tag{1}$$

Assuming that ΔH, the molar heat of fusion of the solvent, is constant over the range of temperature between the freezing point T of the solution and the freezing point T_0 of the pure solvent and integrating equation (1) between these limits

$$\ln a_1 = -\frac{\Delta H}{R}\left(\frac{1}{T} - \frac{1}{T_0}\right) = -\frac{\Delta H}{R}\left(\frac{T_0 - T}{TT_0}\right) \tag{2}$$

Putting $TT_0 \simeq T_0^2$ and $(T_0 - T) = \Delta T$, on substitution in equation (2) we have

$$\ln a_1 = -\frac{\Delta H}{RT_0^2} \cdot \Delta T \tag{3}$$

If a_2 is the activity of the solute at the freezing point of the solution in which there are n_1 moles of solvent and n_2 moles of solute, from the Gibbs–Duhem relation we have

$$d \ln a_2 = -\frac{n_1}{n_2} d \ln a_1 \tag{4}$$

Substituting from equation (3) into equation (4)

$$d \ln a_2 = \frac{n_1}{n_2} \frac{\Delta H}{RT_0^2} d(\Delta T)$$

or

$$d \ln a_2 = \frac{d(\Delta T)}{mK_f} \tag{5}$$

where m is the molality of the solution and K_f is the cryoscopic constant of the solvent.

It is now convenient to define a function j such that

$$j = 1 - \frac{\Delta T}{mK_f} \tag{6}$$

Differentiating equation (6)

$$dj = \frac{\Delta T \, dm}{mK_f^2} - \frac{d(\Delta T)}{mK_f}$$

or

$$\frac{d(\Delta T)}{mK_f} = (1-j)\frac{dm}{m} - dj$$

Hence from equation (5)

$$d \ln a_2 = (1-j)\frac{dm}{m} - dj \tag{7}$$

Subtracting $d \ln m$ from both sides of equation (7)

$$d \ln \frac{a_2}{m} = -j\frac{dm}{m} - dj \tag{8}$$

Integration of equation (8) gives

$$\ln \frac{a_2}{m} = - \int_0^m \frac{j}{m} \, dm - j \tag{9}$$

The integral may be evaluated by measuring the area under the curve of $-j/m$ plotted against m. The activity of the solute may thus be calculated from freezing point data.

It is difficult to obtain accurate data in very dilute solutions but there exists an empirical relationship between j and m. Within the limits of experimental error j/m appears to be constant. Thus if we perform the integration in equation (9) considering j/m as a constant we have

$$\ln \frac{a_2}{m} = - \frac{j}{m} \cdot m - j$$

or

$$2 \cdot 303 \log \frac{a_2}{m} = - 2j \tag{10}$$

Furthermore, a plot of j/m against m should give a horizontal line. If this graph is plotted, the best values of j for various concentrations may be deduced and substituted in equation (10) for the calculation of the activity of the solute.

Apparatus and Chemicals

Beckmann freezing point apparatus, Beckmann thermometer and mannitol.

Method

The Beckmann thermometer is adjusted so that a reading near the top of the scale corresponds to 0°C. A known weight of water is added to the centre tube of the apparatus and the freezing point of the water as indicated by the Beckmann thermometer is determined using the technique described in Experiment 9.

A known weight of mannitol is added to the water to give an approximately 0·1 M solution. The freezing point of this solution is determined.

Further amounts of mannitol are added to give approximately 0·2, 0·3, 0·4 and 0·5 M solutions, the freezing points of the solutions being determined at each concentration.

The factor j is calculated for each concentration and a graph of j/m against m is plotted. From the graph the best values of j for each concentration are determined. The activity of the mannitol at each concentration is then calculated from equation (10).

If preferred, the technique described in Experiment 59 may be employed in this experiment.

Activity of an Electrolyte by Cryoscopy

Discussion

If a solution is sufficiently dilute that the heat of dilution may be neglected, the activity of the solvent a_1 at the freezing point of the solution T is given by

$$\left[\frac{\partial \ln a_1}{\partial T}\right]_p = \frac{\Delta H}{RT^2} \tag{1}$$

where ΔH is the molar heat of fusion of the pure solvent at the freezing point of the solution.

ΔH is dependent upon temperature and can be expressed in terms of temperature by the Kirchhoff equation

$$\Delta H = \Delta H_0 - \Delta C_p \theta \tag{2}$$

where ΔH_0 is the molar heat of fusion of the solvent at the freezing point of the pure solvent, θ is the depression of the freezing point, and ΔC_p is the difference in the heat capacities at constant pressure of the liquid and the solid solvent.

Substituting from equation (2) into equation (1)

$$\left[\frac{\partial \ln a_1}{\partial T}\right]_p = \frac{\Delta H_0 - \Delta C_p \theta}{RT^2} \tag{3}$$

If T_0 is the freezing point of the pure solvent then $T = T_0 - \theta$ and $dT = -d\theta$. Substituting in equation (3)

$$-d \ln a_1 = \frac{\Delta H_0 - \Delta C_p \theta}{R(T_0 - \theta)^2} d\theta \tag{4}$$

Integrating equation (4) and neglecting terms involving θ^3 and higher powers of θ gives the result

$$-\ln a_1 = \frac{1}{RT_0^2}\left[\Delta H_0 \theta + \left(\frac{\Delta H_0}{T_0} - \frac{\Delta C_p}{2}\right)\theta^2\right] \tag{5}$$

With water as solvent and inserting the values $\Delta H_0 = 1438$ cal mole^{-1}, $\Delta C_p = 9$ cal °C^{-1} mole^{-1} and $T_0 = 273 \cdot 16$ °K into equation (5) we have

$$-\ln a_1 = 9 \cdot 702 \times 10^{-3}\theta + 5 \cdot 2 \times 10^{-6} \theta^2 \tag{6}$$

Measurements of the depression of freezing point can thus give the activity of the solvent.

If the solution were ideal, the activity of the solvent a_1 would be equal to the mole fraction of the solvent x_1. For non-ideal solutions this is not true and it is convenient to introduce the rational osmotic coefficient g such that

$$a_1 = x_1{}^g \tag{7}$$

Taking logarithms of equation (7) we have

$$\ln a_1 = g \ln x_1$$

or

$$\ln a_1 = g \ln(1 - x_2) \tag{8}$$

where x_2 is the mole fraction of solute in the solution. If the solution is dilute, i.e. if x_2 is very small and taking the first term only of the expansion series

$$\ln(1 - x_2) = -x_2 + x_2{}^2/2 \ldots$$

then equation (8) becomes

$$\ln a_1 \simeq -g x_2 \tag{9}$$

If M_1 is the molecular weight of the solvent and the solute is an electrolyte of molality m which provides ν ions in solution

$$x_2 = \frac{\nu m M_1}{\nu m M_1 + 1000} \tag{10}$$

Once again if the solution is dilute equation (10) becomes

$$x_2 \simeq \frac{\nu m M_1}{1000} \tag{11}$$

Substituting from equation (11) into equation (9)

$$\ln a_1 \simeq -g \cdot \frac{\nu m M_1}{1000} \tag{12}$$

Equation (12) is an approximate relationship and a practical osmotic coefficient ϕ is defined such that

$$\phi = -\frac{1000}{\nu m M_1} g \ln x_1$$

hence

$$g \ln x_1 = -\frac{\phi \nu m M_1}{1000} \tag{13}$$

and substituting from equation (13) into equation (8) we have the exact relationship

$$\ln a_1 = -\frac{\phi v m M_1}{1000} \tag{14}$$

With water as solvent, putting $M_1 = 18$, equation (14) takes the form

$$\ln a_1 = -\frac{\phi v m}{55 \cdot 51} \tag{15}$$

Substituting the value of $\ln a_1$ given by equation (6) we have

$$\phi = \frac{1}{vm}(0 \cdot 5382\theta + 0 \cdot 00028\theta^2) \tag{16}$$

The importance of the practical osmotic coefficient is that it may be used to calculate the activity coefficient of the solute. From the Gibbs–Duhem relation we have

$$n_1 \, d \ln a_1 + n_2 \, d \ln a_{\pm} = 0 \tag{17}$$

where $a_{\pm}$ is the mean ionic activity of the solute. Putting

$$n_1 = \frac{1000}{M_1} \quad \text{and} \quad n_2 = vm$$

and substituting in equation (17)

$$d \ln a_1 = -\frac{vm M_1}{1000} \, d \ln a_{\pm} \tag{18}$$

Differentiating equation (14) we have

$$d \ln a_1 = -\frac{v M_1}{1000}(\phi \, dm + m \, d\phi) \tag{19}$$

Thus from equations (18) and (19)

$$m \, d \ln a_{\pm} = \phi \, dm + m \, d\phi \tag{20}$$

Now, the mean ionic activity of an electrolyte is related to the mean ionic activity coefficient $\gamma_{\pm}$ and the mean ionic molality $m_{\pm}$ by the equation

$$a_{\pm} = m_{\pm}\gamma_{\pm}$$

hence

$$\ln \gamma_{\pm} = \ln a_{\pm} - \ln m_{\pm} \tag{21}$$

Moreover, the mean ionic molality of an electrolyte which produces v_+ cations and v_- anions in solution is related to the molality m of the solution by the equation

$$m_{\pm}^{\,v} = m^v(v_+^{\,v_+} v_-^{\,v_-})$$

where $v = v_+ + v_-$.

Substituting in equation (21)

$$\ln \gamma_{\pm} = \ln a_{\pm} - \ln m - \ln(\nu_{+}\nu_{-})^{1/\nu} \tag{22}$$

Differentiating

$$d \ln \gamma_{\pm} = d \ln a_{\pm} - d \ln m \tag{23}$$

Substituting the value of $d \ln a_{\pm}$ given by equation (20) into equation (23) we have

$$d \ln \gamma_{\pm} = \phi \frac{dm}{m} + d\phi - d \ln m$$

or

$$d \ln \gamma_{\pm} = (\phi - 1) d \ln m + d\phi \tag{24}$$

Integrating equation (24) from zero molality to molality m and applying the condition that when $m = 0$, $\gamma_{\pm} = \phi = 1$

$$\ln \gamma_{\pm} = \int_{0}^{m} (\phi - 1) d \ln m + (\phi - 1) \tag{25}$$

The integral in equation (25) is evaluated graphically and this can best be done by changing the variable to $\sqrt{m}$ so that

$$\ln \gamma_{\pm} = 2 \int_{0}^{m^{1/2}} \frac{(\phi - 1)}{\sqrt{m}} dm^{1/2} + (\phi - 1) \tag{26}$$

For measuring the freezing points of the solutions, thermistors may be employed instead of a Beckmann thermometer. Thermistors have advantages over a Beckmann thermometer in that they do not require setting, no emergent stem corrections have to be made and, with a suitable bridge, they can be more sensitive and accurate than Beckmann thermometers.

An improvement in sensitivity is achieved if two thermistors are immersed in the solution, the thermistors comprising opposite arms of a Wheatstone bridge circuit as illustrated in Fig. 1. In the diagram the two thermistors are represented by Th_1 and Th_2, R_f is a fixed resistance, R_v a variable resistance, and R_s a resistance of about 20 $k\Omega$ to limit the self heating of the thermistors. The supply to the bridge is provided by a 2 volt accumulator. The resistances, Th_1 and Th_2, of the thermistors are related to the absolute temperature T by the equations

$$Th_1 = A_1 e^{b_1/T} \qquad Th_2 = A_2 e^{b_2/T} \tag{27}$$

where A_1, b_1, A_2 and b_2 are constants. When the bridge is balanced

$$\frac{Th_1}{R_f} = \frac{R_v}{Th_2}$$

or

$$Th_1 . Th_2 = R_v R_f$$

F*

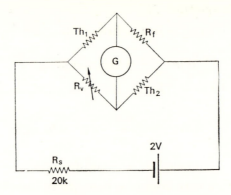

FIG. 1. Wheatstone bridge circuit.

Substituting from equations (27)

$$A_1 e^{b_1/T} . A_2 e^{b_2/T} = R_v R_f$$

or

$$A_1 A_2 e^{(b_1+b_2)/T} = R_v R_f$$

This relation may be written

$$R_v R_f = A' e^{b'/T} \qquad (28)$$

where

$$A' = A_1 A_2 \quad \text{and} \quad b' = b_1 + b_2$$

If the freezing point of the pure solvent is T_0 and the corresponding value of R_v is R_0, then

$$R_0 R_f = A' e^{b'/T_0} \qquad (29)$$

Similarly, denoting the freezing point of a solution as T and the corresponding value of R_v as R,

$$R R_f = A' e^{b'/T} \qquad (30)$$

Dividing equation (30) by equation (29)

$$\frac{R}{R_0} = e^{b'(1/T - 1/T_0)}$$

or

$$\frac{R}{R_0} = e^{b'\{(T_0 - T)/(T T_0)\}}$$

Putting $(T_0 - T) = \theta$ gives

$$\frac{R}{R_0} = e^{b'\theta / \{T_0(T_0 - \theta)\}}$$

and replacing R by $(R_0 + \Delta R)$

$$\frac{R_0 + \Delta R}{R_0} = e^{b'\theta / \{T_0(T_0 - \theta)\}}$$

Taking logarithms

$$\ln\left(1 + \frac{\Delta R}{R_0}\right) = \frac{b'\theta}{T_0(T_0 - \theta)} \tag{31}$$

The logarithmic term in equation (31) may be expanded to as many terms as are necessary for accuracy

$$\left[\left(\frac{\Delta R}{R_0}\right) - \frac{1}{2}\left(\frac{\Delta R}{R_0}\right)^2 + \frac{1}{3}\left(\frac{\Delta R}{R_0}\right)^3 + \ldots\right] = \frac{b'\theta}{T_0(T_0 - \theta)}$$

from which

$$\theta = \frac{T_0^2}{b' + T_0\left[\left(\frac{\Delta R}{R_0}\right) - \frac{1}{2}\left(\frac{\Delta R}{R_0}\right)^2 + \frac{1}{3}\left(\frac{\Delta R}{R_0}\right)^3 + \ldots\right]} \tag{32}$$

Apparatus and Chemicals

Dewar flask, two STC type F23 thermistors, stirrer, frozen distilled water, potassium chloride, 0·05 M silver nitrate solution.

Method

The thermistors are best located in tubes containing a little mercury. These tubes are carried in the stopper of the Dewar flask as illustrated in Fig. 2.

Some frozen distilled water is broken up and the Dewar flask is half filled with it. Enough distilled water (previously cooled to 0°C) is added to just cover the ice. The stopper of the flask is placed in position so that the tubes carrying the thermistors are immersed in the liquid. The solution is gently stirred to facilitate the establishment of equilibrium, the bridge is balanced and the resistance R_0 is noted.

Some concentrated potassium chloride solution (previously cooled to 0°C) is now added so that the solution will be roughly 0·01 M with respect to potassium chloride. The bridge is again balanced and the resistance R is noted. A sample of the solution (about 100 ml will be required at this concentration) is taken and titrated with 0·05 M silver nitrate using potassium chromate as an indicator. The concentration of the potassium chloride solution is calculated.

More potassium chloride is added and the process is repeated. Further additions of concentrated potassium chloride solution should be made until a depression of about 1°C is obtained. Concentrations of potassium chloride of 0·01, 0·02, 0·05, 0·1, 0·2 and 0·3 M should be suitable.

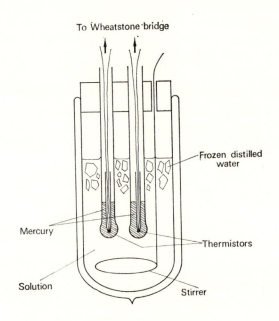

To Wheatstone-bridge

Frozen distilled water

Mercury

Thermistors

Solution

Stirrer

FIG. 2. Freezing point apparatus.

The freezing point depressions of the solutions are calculated from equation (31) and the osmotic coefficients are calculated using equation (16), taking $\nu = 2$ for potassium chloride.

In dilute solutions there will be a considerable experimental error in ϕ but it is known that as $m \to 0$, $\phi \to 1$. Also the Debye–Hückel limiting law will apply in extremely dilute solutions and in the case of a uni-univalent electrolyte at 0°C it takes the form

$$\phi = 1 - 0\cdot374\sqrt{m}$$

Thus, in a graph of ϕ against $\sqrt{m}$, as $m \to 0$, $\phi \to 1$ and the graph should also be asymptotic to the line $\phi = 1 - 0\cdot374\sqrt{m}$ at low concentrations. By plotting this graph more accurate values of ϕ for the various solutions may be deduced.

These improved values of ϕ are used to calculate $\gamma_\pm$ from equation (26). The value of the integral in equation (26) is determined graphically by measuring the area under the curve in a plot of $(\phi-1)/\sqrt{m}$ against $\sqrt{m}$.

The thermistor pair must be calibrated in order to determine the value of b' in equation (32). The resistance R_v must be measured at two known temperatures, preferably in the region of the freezing points of the solutions used in the experiment. Since, in the course of the experiment, the resistance R_v will be measured when the thermistors are at the freezing point of pure water (273·15°K) only one other known temperature is required for calibration.

Activity Coefficient by an e.m.f. Method

Discussion

The e.m.f., E, of the cell

$$\text{Cd–Hg/Cd Cl}_2(m),\ \text{AgCl}(s)/\text{Ag}$$

is given by

$$E = E_{\text{AgCl}} - E_{\text{Cd}}$$

where E_{AgCl} and E_{Cd} are the potentials of the silver–silver chloride and cadmium amalgam electrodes respectively. Further,

$$E_{\text{AgCl}} = E^0_{\text{AgCl}} - \frac{RT}{F}\ln a_-$$

and

$$E_{\text{Cd}} = E^0_{\text{Cd}} + \frac{RT}{2F}\ln\frac{a_+}{a_{\text{M}}}$$

where a_- is the activity of the chloride ions, a_+ is the activity of the cadmium ions and a_{M} is the activity of the cadmium metal in the amalgam. Hence

$$E = (E^0_{\text{AgCl}} - E^0_{\text{Cd}}) - \frac{RT}{F}\ln a_- - \frac{RT}{2F}\ln\frac{a_+}{a_{\text{M}}}$$

or

$$E = E^0 - \frac{RT}{2F}\ln a_-^2 - \frac{RT}{2F}\ln a_+ + \frac{RT}{2F}\ln a_{\text{M}} \tag{1}$$

where E^0 is the standard e.m.f. of the cell. If the activity of the cadmium in the amalgam remains constant, the first and last terms on the right-hand side of equation (1) may be combined into a constant E' so that

$$E = E' - \frac{RT}{2F}\ln a_-^2 - \frac{RT}{2F}\ln a_+$$

or

$$E = E' - \frac{RT}{2F}\ln a_+ a_-^2 \tag{2}$$

It is not possible to determine the activity of a single ionic species but only a mean ionic activity $a_\pm$ for an electrolye. For an electrolyte which gives ν_+ cations and ν_- anions, the mean ionic activity is defined in terms of the individual ionic activities by

$$(a_\pm)^\nu = (a_+)^{\nu_+} \cdot (a_-)^{\nu_-}$$

where

$$\nu = \nu_+ + \nu_-$$

Thus, for cadmium chloride

$$(a_\pm)^3 = (a_+) \cdot (a_-)^2$$

and substituting in equation (2)

$$E = E' - \frac{RT}{2F} \ln(a_\pm)^3$$

or

$$E = E' - \frac{3RT}{2F} \ln a_\pm \tag{3}$$

The mean ionic activity $a_\pm$ is related to the mean ionic molality $m_\pm$ and the mean ionic activity coefficient $\gamma_\pm$ by

$$a_\pm = m_\pm \gamma_\pm \tag{4}$$

where

$$(m_\pm)^3 = (m_+) \cdot (m_-)^2 \tag{5}$$

Substituting from equation (4) into equation (3) gives

$$E = E' - \frac{3RT}{2F} \ln m_\pm \gamma_\pm \tag{6}$$

If the e.m.f. of the cell is determined at various molalities of cadmium chloride and the constant E' is known it is possible to calculate the mean ionic activity coefficients for cadmium chloride at the various molalities studied by using equation (6).

The constant E' may be found from the experimental results in the following way. Rearranging equation (6)

$$E + \frac{3RT}{2F} \ln \gamma_\pm = E' - \frac{3RT}{2F} \ln m_\pm$$

or

$$E + \frac{6 \cdot 909 RT}{2F} \log \gamma_\pm = E' - \frac{2 \cdot 303 RT}{2F} \log(m_\pm)^3 \tag{7}$$

The activity coefficient term may be expressed as

$$\log \gamma_\pm = -0{\cdot}5 z_+ z_- \sqrt{I} + CI \tag{8}$$

where z_+ and z_- are the number of charges carried by the cation and anion respectively and C is a constant.

Substituting from equation (8) into equation (7)

$$E - \frac{6{\cdot}909RT}{2F}\sqrt{I} + \frac{6{\cdot}909RT}{2F}CI = E' - \frac{2{\cdot}303RT}{2F}\log(m_\pm)^3$$

For cadmium chloride solution of molality m,

$$m_+ = m \quad \text{and} \quad m_- = 2m$$

Hence, from equation (5)

$$(m_\pm)^3 = 4m^3 \tag{9}$$

The ionic strength I of the solution is given by

$$I = \tfrac{1}{2}\sum (c_i z_i^2)$$

and neglecting the difference between molality and molarity, the ionic strength of a cadmium chloride solution of molality m is given by

$$I = 3m \tag{10}$$

Hence

$$E - \frac{6{\cdot}909RT}{2F}\sqrt{3m} + \frac{6{\cdot}909RT}{2F}3Cm = E' - \frac{2{\cdot}303RT}{2F}\log(4m^3)$$

Rearranging

$$E - \frac{6{\cdot}909RT}{2F}\sqrt{3m} + \frac{2{\cdot}303RT}{2F}\log(4m^3) = E' - \frac{6{\cdot}909RT}{2F}3Cm \tag{11}$$

A plot of the left-hand side of equation (11) against m should, at low concentrations give a straight line of intercept E'.

Apparatus and Chemicals

Cadmium electrode, mercury, silver–silver chloride electrode, potentiometer and accessories, $0{\cdot}1$ M cadmium chloride solution.

Method

The cadmium amalgam electrode is prepared by rubbing the surface of the cadmium rod with a little mercury under dilute hydrochloric acid. The silver–silver chloride electrode is prepared as described in Experiment 51. Both electrodes are well washed with distilled water and then with $0{\cdot}005$ M cadmium chloride solution made by dilution of the stock solution. A cell is assembled by immersing the electrodes in a 0·005 M solution of cadmium chloride. The e.m.f. of the cell is measured with a potentiometer.

Further cells are set up containing cadmium chloride of different concentrations, for example, 0·0075, 0·010, 0·025, 0·050, 0·075 and 0·100 M. The e.m.f.'s of these cells are measured.

The experiments are best carried out in a thermostat but if the conditions are such that the temperatures of the various cells are within 1–2°C of each other the work may be done on the bench.

The quantity on the left-hand side of equation (11) is plotted against m to determine E'. From the result, $\gamma_\pm$ for the various concentrations of cadmium chloride is calculated using equation (6). From the graph the value of the constant C is determined and comment should be made on its value.

The Partial Molal Volumes of a Binary Solution

Discussion

The value of any extensive property of a homogeneous system, e.g. the volume, the entropy, or the free energy, is determined by the state of the system and by the amount of the substance present. If F is any extensive property, then

$$F = f(T, P, n_1, n_2, \ldots n_i \ldots)$$

where n_i is the number of moles of the ith component of the system. On differentiating

$$dF = \left(\frac{\partial F}{\partial T}\right)_{P, n_1, n_i} \cdot dT + \left(\frac{\partial F}{\partial P}\right)_{T \, n_1, n_i} dP.$$

$$+ \left(\frac{\partial F}{\partial n_1}\right)_{T, P, n_i} \cdot dn_1 \ldots + \left(\frac{\partial F}{\partial n_i}\right)_{T, P, n_1} \cdot dn_i$$

The derivative

$$\left(\frac{\partial F}{\partial n_i}\right)_{T, P, n_i}$$

is the partial molal property of the constituent i. It is an intensive property and is written $\bar{F}_i$.

At constant temperature and pressure:

$$dF_{T,P} = \bar{F}_1 \cdot dn_1 + \ldots + \bar{F}_i \cdot dn_i + \ldots$$

hence

$$F_{T,P} = \bar{F}_1 n_1 + \ldots + \bar{F}_i \cdot n_i + \ldots$$

A partial molal property suitable for experimental study is the partial molal volume. For a binary solution consisting of n_1 moles of solvent and n_2 moles of solute, the partial molal volumes of the solvent $\bar{V}_1$ and of the solute $\bar{V}_2$, are related to the volume V thus:

$$V = \bar{V}_1 n_1 + \bar{V}_2 n_2$$

The simplest way of determining the partial molal volume of the solute would be to plot the volume of solution for varying concentrations of the solute against the molality and to determine the slope $(\partial V/\partial n_2)_{T,P}$. This is not a very good method; the actual volumes of different concentrations would be difficult to determine accurately. Instead, an approximate function may be used, and the difference between this function and the volume evaluated at each concentration.

The apparent molal volume of the solute ϕ_2 is defined by the relationship

$$\phi_2 = \frac{V - n_1 \cdot V_0}{n_2} \tag{1}$$

where V_0 is the molar volume of the pure solvent. This function can be used to determine the partial molal volumes. Rearranging this equation

$$V = n_2 \cdot \phi_2 + n_1 \cdot V_0$$

and

$$\bar{V}_2 = \left(\frac{\partial V}{\partial n_2}\right)_{T,P,n_1} = \phi_2 + n_2 \cdot \left(\frac{\partial \phi_2}{\partial n_2}\right)_{n_1,T\ P} \tag{2}$$

The partial molal volume of the solvent is given by

$$\bar{V}_1 = \frac{V - n_2 \bar{V}_2}{n_1} = 1/n_1 \cdot \left\{ n_1 V_0 - n_2{}^2 \cdot \left(\frac{\partial \phi_2}{\partial n_2}\right)_{n_1,T\ P} \right\} \tag{3}$$

The apparent molal volume of the solute can be calculated from the density of the solution. If M_1 and M_2 are the molecular weights of the solvent and solute, respectively, and d is the density of the solution; then ϕ_2 as defined by equation (1) is given by

$$\phi_2 = 1/n_2 \left\{ \frac{n_1 M_1 + n_2 M_2}{d} - n_1 \cdot V_0 \right\}$$

If n_1 is equal to the number of moles of the solvent and n_2 the number of moles of solute in 1000 g of the solvent, then $n_2 = m$, the molality and

$$\phi_2 = 1/m \left\{ \frac{1000 + m M_2}{d} - 1000/d_0 \right\} = \left\{ \frac{1000}{m d d_0} (d_0 - d) + M_2/d \right\} \tag{4}$$

where d_0 is the density of the pure solvent. In this way the apparent molal volume of the solute can be determined for a molality m.

In order to determine $(\partial \phi_2/\partial n_2)_{n_1,T,P}$ a plot of apparent molal volumes (ϕ_2) vs. m should be made and the slope determined. In the case of electrolytes it has been found that the apparent molar volume increases linearly with the square root of the concentration, and in dilute aqueous solution may be written

$$\phi_2 = \varphi_0 + a\sqrt{m}$$

where ϕ_0 is the apparent molal volume at infinite dilution and a is a constant. Hence, instead of determining $(\partial\phi_2/\partial n_2)_{n_1,T,P}$ from a plot of ϕ_2 vs. m we plot ϕ_2 vs. $m^{1/2}$ and obtain $(\partial\phi_2/\partial m^{1/2})_{n_1,T,P}$ from the slope and hence calculate $(\partial\phi_2/\partial m)_{n_1,T,P}$ from the relationship

$$\left(\frac{\partial\phi_2}{\partial m}\right)_{n_1,T\ P} = \frac{1}{2m^{1/2}}\left[\left(\frac{\partial\phi_2}{\partial(m^{1/2})}\right)_{n_1,T,P}\right] \tag{5}$$

Apparatus and Chemicals

Pyknometers, five small glass stoppered bottles and sodium chloride.

Method

Five solutions of sodium chloride in water are prepared containing 2, 4, 8, 12 and 16 g of the solute per 100 g of water. (As the density of water is 0·9982 g ml^{-1} at 20°C, the temperature at which pipettes are calibrated, 100 g of water at 20°C equals 100·2 ml.)

The density of each solution at 25°C is determined accurately.

The apparent molal volume of the solute ϕ_2 for each solution is calculated using equation (4).

A plot of ϕ_2 vs. $m^{1/2}$ where m is the molality of the solution is drawn and the slope $(\partial\phi_2/\partial m^{1/2})_{n_1,T,P}$ determined. The value of $(\partial\phi_2/\partial m)_{n_1,T,P}$ is calculated using equation (5).

The partial molal volumes of the solvent $\bar{V}_1$, and of the solute $\bar{V}_2$ are calculated using equations (3) and (2), respectively.

A Ternary Phase Diagram for a System of Two Solids and a Liquid[1]

Discussion

The isothermal phase diagram for the solid–liquid ternary system water–sodium nitrate–lead nitrate is of the form illustrated in Fig. 1.

The symbols a, b, c refer to the percentage by weight of water, sodium nitrate and lead nitrate severally. a_s, b_s and c_s are the values of a, b and c on the solubility curve.

Values of a_s vs. $b/(b+c)$ are plotted for two series of synthetic samples. In order to ensure the presence of at least two phases, samples are chosen which on visual examination indicate a liquid phase and at least one solid phase. In each series the water concentrations a_1 and a_2 are kept constant. The two graphs so obtained correspond to the lines *lmnd* and *lrqd* in Fig. 2.

Two synthetic samples $a_1b_1c_1$ and $a_2b_2c_2$ which have the same a_s value, as determined from Fig. 2, must lie on the same tie-line (Fig. 1) and that tie-line must also be common to the point $a_sb_sc_s$.

The tie-lines and solubility curve for this system can hence be constructed since by application of the theorems of similar triangles

$$\frac{a_s - a_1}{a_1 - a_2} = \frac{b_s - b_1}{b_1 - b_2} = \frac{c_s - c_1}{c_1 - c_2} \tag{1}$$

From these two equations the only two unknowns c_s and b_s may be obtained.

The phase rule for a ternary system at constant temperature and pressure reduces to

$$F = 3 - P$$

Thus there is an invariant point ($F = 0$) at the intersection of the two branches of the solubility curve corresponding to samples containing three phases ($P = 3$).

The composition of such samples is represented in Fig. 2 by the line *rmnq* since in this region a_s does not vary with $b/(b+c)$.

167

Apparatus and Chemicals

Thermostat at 25°C, seventeen 60 ml glass stoppered bottles, seventeen 50 ml conical flasks, seventeen 5 ml container pipettes, 10 ml pipette, pure samples of sodium nitrate and lead nitrate.

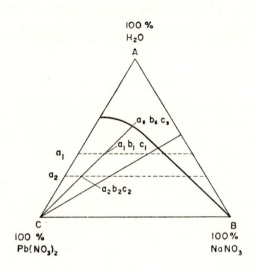

Fig. 1. Phase diagram for the isothermal ternary system water–sodium nitrate–lead nitrate.

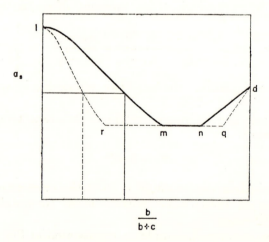

Fig. 2. Graphs of a_s vs. $b/(b+c)$ for two series of synthetic samples.

Method

The bottles, conical flasks and pipettes are cleaned and dried. Ten millilitres of water, delivered from the 10 ml pipette, are accurately weighed. The two series of synthetic samples are then prepared in the glass-stoppered bottles. The quantities for each sample are as shown in Table 1.

TABLE 1

Bottle No.	Water (ml)	Sodium nitrate (g)	Lead nitrate (g)	
1	10	15		
2	10		15	
3	10	14·250	0·750	
4	10	13·000	2·000	
5	10	12·000	3·000	$[H_2O] = a_1$
6	10	8·000	7·000	
7	10	4·005	11·000	
8	10	1·505	13·500	
9	10	2·280	27·725	
10	10	5·650	24·360	
11	10	22·000	8·000	
12	10	17·000	13·000	
13	10	14·000	16·000	$[H_2O] = a_2$
14	10	28·875	1·120	
15	10	27·720	2·280	
16	10	27·000	3·000	
17	10	8·000	22·005	

The glass-stoppered bottles are stored in a thermostat at 25°C for 2 hr and the contents swirled at frequent intervals.

An approximately 2 ml fraction of solution is removed from one of the samples. A container pipette that is well above 25°C is used for this purpose to prevent solid crystallizing out. The fraction is drained into a weighed conical flask. The stopper is replaced and the flask and contents weighed. This procedure is repeated for all of the samples using a clean dry conical flask and pipette for each sample.

The flasks containing the fractions are heated to less than 130°C to remove the water, stoppered, cooled in a desiccator and reweighed. The losses in weight on heating expressed as percentages are the a_s values.

Graphs are plotted of a_s vs. $b/(b+c)$ for the two constant values of a, i.e. a_1 and a_2.

The $a_1b_1c_1$ and the corresponding $a_2b_2c_2$ for a number of a_s values may be obtained from the graph. These results may then be substituted in equation (1) in order to determine b_s and c_s values.

It should be noted that since $(a+b+c)$ is equal to 100 per cent, $(b+c)$ is known for any a value. Hence knowing the fraction $b/(b+c)$ the separate b and c values can be determined.

The series of $a_s b_s c_s$ values, when plotted on a triangular diagram, form the solubility curve for this ternary system.

[1] E. L. HERIC, *J. Chem. Educ.*, **35**, 510–13 (1958).

Emission Spectra Study of Atomic Hydrogen

Discussion

The hydrogen atom is the simplest known. It consists of a proton, of charge $+e$ and mass M, and an electron, of charge $-e$ and mass m. It is possible to solve the wave equation for a simple system of this type and it can be shown that the allowed values E of the energy levels are given by

$$E = -\frac{2\pi^2 m e^4}{n^2 h^2} \tag{1}$$

where the principal quantum number $n = 1, 2, 3, ...,$ and h is Planck's constant.

When a hydrogen atom loses energy ΔE by allowing electrons to fall from a high energy level to a lower one, it gives out light, the frequency of which is governed by the equation

$$\Delta E = h\nu \tag{2}$$

Combining equations (1) and (2), the frequency of the light emitted is given by the relationship

$$\nu = -\frac{2\pi^2 m e^4}{h^3} \left(\frac{1}{n_2^2} - \frac{1}{n_1^2} \right) \tag{3}$$

Hence the wavenumber $\bar{\nu}$ is given by the equation

$$\bar{\nu} = \frac{2\pi^2 m e^4}{h^3 c} \left(\frac{1}{n_1^2} - \frac{1}{n_2^2} \right) \tag{4}$$

As the wavenumber is equal to the reciprocal of the wavelength, equation (4) may be rewritten

$$\frac{1}{\lambda} = \frac{2\pi^2 m e^4}{h^3 c} \left(\frac{1}{n_1^2} - \frac{1}{n_2^2} \right) \tag{5}$$

Equation (5) corresponds to the empirical equation

$$\frac{1}{\lambda} = R \left(\frac{1}{n_1^2} - \frac{1}{n_2^2} \right) \tag{6}$$

where R is a constant known as the Rydberg constant and

$$n_2 = (n_1 + 1), (n_1 + 2), (n_1 + 3) \ldots.$$

From equations (5) and (6) we can see that

$$R = \frac{2\pi^2 m e^4}{h^3 c} \tag{7}$$

Several spectral series occur in the hydrogen spectrum, corresponding to different integral values of n. In the Balmer series $n_1 = 2$ and hence $n_2 = 3, 4, 5, \ldots$. If the wavelengths of the Balmer lines are determined it is possible to deduce a value for the Rydberg constant from equation (6).

Apparatus and Chemicals
Hilger medium quartz spectrograph, hydrogen discharge lamp, photographic plates, e.g. Ilford N.30, dark room facilities.

Method
The hydrogen discharge lamp is set up so that the capillary portion is about 3 cm from the entrance slit of the spectrograph. The slit width should be set to 0·05 cm. The supply to the hydrogen lamp is turned on and the discharge started. (Caution: As the supply to the discharge is at 5000 V, the lamp must not be touched when running.) The light must completely fill the entrance slit. This may be checked by observing the red α-line from the position of the plate holder. The red α-line is very intense and it is advisable to attenuate it by placing a suitable blue filter in front of the entrance slit. The supply to the discharge is turned off.

The plate holder is loaded in a dark room with a suitable plate, making sure that the emulsion side of the plate will be exposed to the discharge. The emulsion side of the plate is smooth, the rough side is the backing. The holder is replaced in the spectrograph.

The supply to the discharge is turned on and the plate is exposed. It is unlikely that the correct exposure time will be achieved at the first attempt. Several preliminary exposures should be made with the first plate to determine the correct conditions for the second plate. The plate is exposed to the hydrogen spectrum and the internal scale of the instrument.

The plate holder is removed and the plate is unlaced in a dark room where it is developed.

The dry plate is examined with the eye or it may be projected if facilities are available. The wavelengths of the Balmer lines are measured using the internal scale of the instrument. This scale gives wavelengths directly with an error of 0·2 per cent, which is satisfactory for this experiment.

The first five lines of the Balmer series are given in Table 1 (λ_{vac} denotes the wave-length in vacuum):[1]

TABLE 1

Line	λ_{vac} (Å)
1	6564·729
2	4862·761
3	4341·746
4	4102·948
5	3971·246

The lines on the plate are identified with the aid of this table and the value of the Rydberg constant is calculated using equation (6). An energy level diagram is drawn for the hydrogen atom using equations (2) and (3) and the observed transitions are noted.

The theoretical value of the Rydberg constant is calculated from equation (7) taking $e = 4 \cdot 80 \times 10^{-10}$ e.s.u., $m = 9 \cdot 11 \times 10^{-28}$ g, $h = 6 \cdot 62 \times 10^{-27}$ erg sec and $c = 3 \cdot 00 \times 10^{10}$ cm sec^{-1}. The theoretical value for R is compared with the experimental value.

[1] H. C. UREY, F. G. BRICKWEDDE and G. M. MURPHY, *Phys. Rev.* **40**, 1, (1932).

The Ultra-violet Absorption Spectra of Geometrical Isomers

Discussion

Azobenzene usually consists largely of the *trans*-isomer; the corresponding *cis* isomer may be present up to 1 per cent. The amount of the *cis* isomer may be increased by irradiating the mixture with ultra-violet light.

I *trans*   II *cis*

The mixture may be separated by chromatographic analysis.

Chromatography is essentially a process of differential adsorption. In liquid chromatography, the liquid phase (containing, in this case, the mixture of isomers) is allowed to flow over the solid phase (alumina). The solid phase acts as an adsorbent and due to the different degrees of adsorption the isomers separate into different zones. The pattern of zones is called the chromatogram.

The spectra of the isomeric azobenzenes contain two characteristic bands.[1] An intense band occurs in the region of 315 to 325 mμ. This is the K-band due to the conjugation between the N=N group and the aromatic nuclei. A second weaker band occurs in the 440 to 450 mμ region. This is the R-band and is ascribed to the N=N linkage. In the case of the isomeric azobenzenes, a change from *trans* to *cis* involves a change in the intensity for both bands and an appreciable shift in the wavelength. If, in the excited state, both isomers are uniplanar (III), then the spatial arrangement of the isomer in the ground state which favours this coplanar structure, i.e. I, should lead to a greater increase in intensity.

III

174

This is true in the case of the K-band but not of the R-band. In the latter case, presumably, the electronic transition does not involve the movement of the conjugated bond.

Apparatus and Chemicals

Chromatographic column containing alumina as the adsorbent, ultra-violet lamp, spectrophotometer, e.g. Unicam SP.500, glass pestle, azobenzene.

Method

A quantity of azobenzene is dissolved in light petroleum (b.p. 60–80°C) to give a solution containing about 4 g.l^{-1}. The solution is irradiated under an ultraviolet lamp for about 30 min. After the irradiation period it is essential to protect all solutions containing the *cis*-isomer from light and heat. The chromatographic column should therefore be shielded from daylight with a wrapping of black paper. Similarly, flasks, beakers, etc., must be protected.

During the period of irradiation the column for chromatographic analysis should be prepared. A glass tube is plugged at one end with glass wool. The alumina is introduced into the tube, each portion of alumina added being well pressed down with a glass pestle. A column of about 5 cm is sufficient.

Light petroleum is added to the packed tube. The solvent is allowed to flow out until finally a layer of solvent of about 1 cm remains at the top of the column. Twenty millilitres of the irradiated solution of azobenzene is added carefully and allowed to flow through the column at a rate of about 1 to 2 ml per minute. Further quantities of light petroleum are added to the column, taking care not to disturb the top layer. In all, 100 ml of the solvent should be added for complete elution of the *trans*-isomer. The solution containing the *trans*-isomer is collected in a beaker.

The solution is evaporated to dryness. The residue is weighed and a solution in light petroleum of concentration about 0.8×10^{-4} M is prepared. A portion of the solution is placed in a silica cell and the absorbance A, in the region 450–300 mμ, is measured with a spectrophotometer. The molar absorptivity ϵ, is calculated using the formula

$$\epsilon = \frac{A}{cl}$$

where c is the concentration in mole l^{-1}. and l is the length of the cell. The absorption curve ϵ against λ is drawn and λ_{max} and ϵ_{max} are noted for the K- and R-bands.

A further 20 mi of the irradiated solution is evaporated to dryness. The residue obtained gives the total weight of the *cis-* and *trans-*isomers per 20 ml of the original solution, if this is not known definitely. The weight of the *cis-*isomer is therefore found by difference. (Solutions containing the *cis-*isomer only must never be heated as this would cause reversion to the *trans-*form.)

The *cis-*isomer remains as a compact band near the top of the column and can be removed by elution with light petroleum containing 1 per cent methanol. The solution obtained is adjusted to give a final *cis-*concentration of about 0.8×10^{-4} M. The ultraviolet spectrum of the *cis-*solution is obtained using methanol petroleum as a blank. The cell is exposed to sunlight and the time noted. The spectrum is re-determined after 15 min. This operation is repeated at known intervals of time until only the *trans-*isomer spectrum is obtained.

[1] A. H. Cook, D. G. Jones and J. B. Polya, *J. Chem. Soc.*, 1315 (1939).

Force Constants from Vibrational Frequencies

Discussion

The absorption of infra-red radiation by a substance causes an increase in its vibrational and/or rotational energy. With a given electronic energy level there are associated quantized vibrational and rotational energy levels. Transitions between electronic levels, generally requiring greater energy than do transitions between vibrational states, are excited by the visible and ultra-violet frequencies. At sufficiently high resolution, the spectrum corresponding to a transition between two electronic energy levels, is seen to have a fine structure which can be traced to concomitant changes in the vibrational frequencies.

The energy quanta of near infra-red frequencies increase the vibrational energy within the ground electronic level. In a diatomic species the vibrational spectrum can be unambiguously associated with the single vibrational degree of freedom of the molecule; a polyatomic system gives a more complex spectrum, certain frequencies corresponding to transitions between vibrational states of a particular bond, others to transitions between coupled vibrational modes characteristic of larger groups of atoms or the whole molecule.

To a first approximation, two directly bonded atoms in a molecule may be regarded as vibrating with simple harmonic motion, in which case

$$\omega = \frac{1}{2\pi} \sqrt{\frac{f}{\mu}} \tag{1}$$

where ω is the mechanical frequency of oscillation, f the restoring force in dynes cm^{-1} and μ, the reduced mass, in terms of masses of atoms m_1 and m_2 is given by

$$\mu = \frac{m_1 \cdot m_2}{m_1 + m_2}$$

In a quantized state the energy E_v is given by

$$E_v = (v + \tfrac{1}{2})h\omega$$

where h is Planck's constant and v the vibrational quantum number which may have integral values. It is seen that the energy absorbed in a transition from the lowest to the next vibrational energy state will be $h\omega$, and an absorption band of frequency $v = \omega$ will appear in the spectrum. In the absence of strong coupling, and subject to certain selection rules, a band will show its characteristic frequency in the spectrum of any compound in which it occurs. Thus carbonyl compounds show in their infra-red spectra an absorption band at a frequency characteristic of the vibrational stretching mode of the carbonyl double bond. The occurrence of a particular absorption band in the spectra of a vast number of carbonyl compounds, and its absence from non-carbonyl compounds, is strong evidence for the assignment of the frequency to this bond. The force constant for the bond may be determined from the relationship

$$f = 4\pi^2\mu v^2 \qquad (2)$$

replacing the mechanical frequency of oscillation by the observed infra-red absorption frequency. In terms of the wavenumber this equation becomes

$$f = 4\pi^2\mu c^2\bar{v}^2 \qquad (3)$$

where c is the velocity of light in cm sec^{-1}.

Apparatus and Chemicals

Infra-red spectrophotometer, e.g. Unicam SP.200, associated cells, freshly distilled samples of benzaldehyde, acetophenone, benzophenone, cyclohexanone and anisole.

Method

Infra-red spectra are recorded for a series of carbonyl compounds such as benzaldehyde, acetophenone, benzophenone, cyclohexanone and an ether, for example, anisole.

Liquids of low volatility may be examined between sodium chloride windows, and more volatile liquids in liquid cells. Solids should be mulled with liquid paraffin. Sample thicknesses should be chosen so that strong bands transmit between, say, 10 per cent and 50 per cent of the incident energy. For each substance used, the frequencies of strong bands, i.e. bands of less than 50 per cent transmittance are tested.

One band common to all the carbonyl compound spectra, but absent from that of the ether will be found; this band is assigned to the carbonyl stretching mode. The centre of this band for each substance used is recorded and the mean taken.

The value of the force constant for the carbonyl bond is calculated by equations (2) or (3).

Infra-red Spectra of Carbonyl Compounds

Discussion

According to electronic theory, the polarization and double bond character of the carbonyl group are modified by substituents on the carbon atom. In the case of substituents with unshared electron pairs, structures of Type 1 increase the carbon to oxygen bond length and hence lead to a reduction in the vibration frequency of the carbonyl group.[1]

An increase in the electronegativity of —X will reduce the significance of the structures such as II, decreasing the polarity of the carbon to oxygen bond and increasing the vibrational frequency.

The vibrational frequency of the carbonyl group is influenced by factors in addition to those due to substituents attached to the carbon atom, for example, dipolar association III, and hydrogen bonding IV.

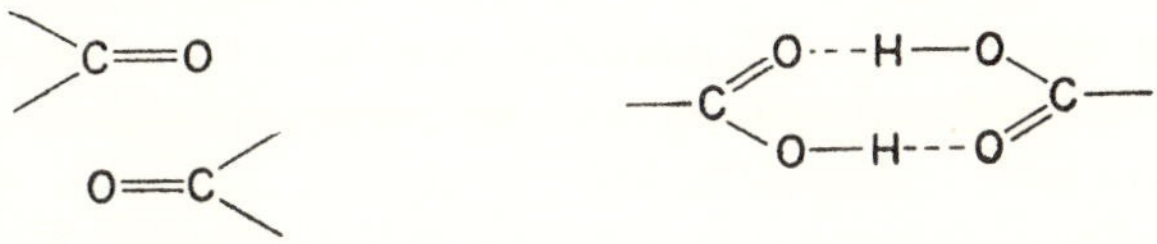

These effects, which reduce the vibration frequency of the carbonyl group, are minimized in dilute solutions of the compound or its vapour, where interactions between the polar structures are smaller. Carboxylic acids in chloroform solution and in the vapour state show two bands for the carbonyl stretching mode: one characteristic of the dimer, the other of the monomer. In the vapour state the intensity of the high-frequency band relative to the low-frequency band increases with rise in temperature. We attribute the higher frequency band to the carbonyl group of the monomer.

G   179

Apparatus and Chemicals

Infra-red spectrophotometer, gas and liquid cells, heating tape, freshly distilled samples of acetic acid, acetyl chloride, acetone and NN-diethylacetamide.

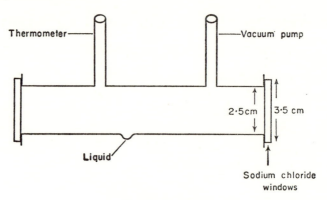

FIG. 1. Gas cell.

Method

Samples of acetic acid, acetyl chloride and acetone are vaporized in the gas cell illustrated in Fig. 1. The spectra of these compounds are obtained at 20°C and 60°C.

Thin films of acetic acid, acetyl chloride and NN-diethylacetamide are placed between sodium chloride plates and the spectra of these compounds recorded.

A liquid cell (0·05 mm thickness) is filled with acetone and the spectrum measured.

Chloroform solutions of NN-diethylacetamide (0·1 M) and acetic acid (0·1 M) are prepared. Each solution in turn is placed in a cell (0·2 mm thickness) and examined in the sample beam of the infra-red spectrophotometer. The cells are separately balanced against pure chloroform contained in a matched cell in the reference beam. Alternatively, if matched cells are not available, the spectra of the chloroform solutions should be compared with the spectrum of chloroform recorded under the same conditions.

The positions of the carbonyl stretching frequencies least affected by hydrogen bonding or dipolar interactions are listed for each spectrum, and the order of frequencies is correlated with the mesomeric and inductive effects of the substituents.

The effect of dipole moment interaction and hydrogen bonding on the vibrational frequency of the carbonyl group can be detected by comparing the spectra of liquid and vapour states.

[1] E. J. HARTWELL, R. E. RICHARDS and H. W. THOMPSON. *J. Chem. Soc.*, 1436 (1948).

Variation of Velocity Constant with Catalyst Concentration by a Dilatometer Method

Discussion

The method described is for determining velocity constants of first-order or pseudo first-order reactions where the equilibrium or final reading for a reaction is not known.[1] For such reactions, if x_1, x_2, x_3, etc., correspond to a suitable physical property measured at time t_1, t_2, t_3, etc., the rate expression may be written

$$(x_1 - x_\infty) = (x_0 - x_\infty) \, e^{-k't_1} \tag{1}$$

where x_0 and x_∞ are the physical measurements for $t = 0$ and $t = \infty$, respectively.

If a second series of readings are made after a constant increment of time Δt such that x_1', x_2', x_3', etc., are the measurements of the physical property made at time $t_1 + \Delta t$, $t_2 + \Delta t$, $t_3 + \Delta t$, etc., then

$$(x_1' - x_\infty) = (x_0 - x_\infty) \, e^{-k'(t_1 + \Delta t)} \tag{2}$$

Subtracting equation (2) from equation (1) gives

$$(x_1 - x_1') = (x_0 - x_\infty) \, e^{-k't_1}(1 - e^{-k'\Delta t}) \tag{3}$$

i.e.

$$k't_1 + \ln(x_1 - x_1') = \ln[(x_0 - x_\infty)(1 - e^{-k'\Delta t})] \tag{4}$$

i.e.

$$k't_1 + \ln(x_1 - x_1') = \text{constant}$$

In general terms this may be written

$$\ln(x_n - x_n') = -k't_n + \text{constant}$$

Hence a graph of $\log(x_n - x_n')$ vs. t_n gives a straight line of slope $-k'/2\cdot303$.

The alkali-catalysed cleavage of diacetone alcohol is a pseudo first-order reaction being dependent on the first power of the diacetone alcohol concentration and on the first power of the hydroxide ion concentration. The hydroxide ion concentration, however, remains constant during reaction. A suitable physical property by which the reaction may be followed is the volume which is readily measured with a dilatometer.

The reaction is investigated at different concentrations of sodium hydroxide and $k'/[NaOH]$ determined.

Apparatus and Chemicals

Two dilatometers (Fig. 1), thermostat at 25°C, cathetometer, metre scale, stop watch, diacetone alcohol, 0·1 N, 0·2 N, and 0·5 N sodium hydroxide solutions.

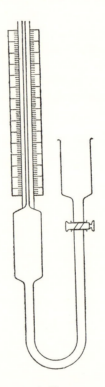

FIG. 1. A dilatometer.

Method

The dilatometer is cleaned, dried and attached to the metre scale. The bulb of the dilatometer is then immersed in a thermostat at 25°C. Thirty-five millilitres of 0·1 N sodium hydroxide is delivered into a conical flask from a burette and the flask placed in the thermostat. After an interval of 15 min, 2·0 ml of diacetone alcohol is added to the flask with the aid of a pipette. The contents of the flask are mixed thoroughly and subsequently poured into the funnel of the dilatometer. The tap is opened and a quantity of the solution is allowed to run into the bulb and the capillary of the dilatometer. When the meniscus of the liquid is at the bottom of the attached scale, the tap is closed. The position of the meniscus, which is observed using the cathetometer, is noted and recorded every 5 min for a period of 30 min. The second series of readings are taken after a time interval of 15 min, i.e. $\Delta t = 45$ min.

During this 15 min interval, the first set of observations are carried out using 0.5 N sodium hydroxide as catalyst. This set of readings should be taken every $\frac{1}{2}$ min for 3 min. For this more concentrated solution of sodium hydroxide the second series of readings are taken after a time interval of 2 min, i.e. $\Delta t = 5$ min.

Since the measured physical property, i.e. the volume, increases during this experiment, equation (1) should be subtracted from equation (2). The results are then recorded in tabular form under the headings:

Concentration NaOH	t_n	x_n	x_n'	$\log(x_n' - x_n)$	k'	$\dfrac{k'}{[\text{NaOH}]}$

Finally a graph of velocity constant against concentration of hydroxyl ion is plotted.[2]

[1] E. A. GUGGENHEIM, *Phil. Mag.*, **2**, 538 (1926).

[2] For a model report of this reaction see A. M. JAMES, *Practical Physical Chemistry*, Churchill (London, 1961), p. 11.

A Velocity Constant by an Amperometric Method[1]

Discussion

The rate expression for the fast reaction

$$C_2H_5NO_2 + OH^- \rightarrow C_2H_4NO_2^- + H_2O$$

is

$$-\frac{d[OH^-]}{dt} = k[C_2H_5NO_2][OH^-] \qquad (1)$$

i.e.

$$k = \frac{-d[OH^-]/dt}{[C_2H_5NO_2][OH^-]} \qquad (2)$$

To investigate the kinetics of this reaction a current is passed through an aqueous solution of potassium chloride and nitroethane until the pH value of the solution remains constant. At this point a steady state is achieved since the rate of production of hydroxyl ions by electrolysis is equal to the rate of hydroxyl ion consumption by the reaction.

From the laws of electrolysis

$$\frac{d[OH^-]}{dt} = \frac{I}{Fv} \qquad (3)$$

where I is the steady state current and v the volume of the catholyte solution in litres.

The hydroxyl ion concentration at the steady state may be determined from the steady state pH value by the relationship

$$\log[OH^-] = pH - 14 \qquad (4)$$

The concentration of nitroethane is considered to remain constant since it is present in a large excess throughout the reaction.

The rate constant for this reaction can now be calculated since all the quantities on the right-hand side of equation (2) can be determined.

Apparatus and Chemicals

Pye pH meter and electrodes (glass and calomel), milliammeter, variable resistance, cylindrical platinum gauze cathode, platinum wire sealed in mercury-filled tube, cation exchange resin Dowex

AG 50 WX8,[2] cation exchange membrane Nepton CR-61,[3] magnetic stirrer, glass tube 12 cm by 1·5 cm, rubber sleeve for glass tube, 400 ml beaker, 4 M sodium chloride, 0·5 M sodium hydroxide, 1 M potassium chloride, potassium chloride and nitroethane.

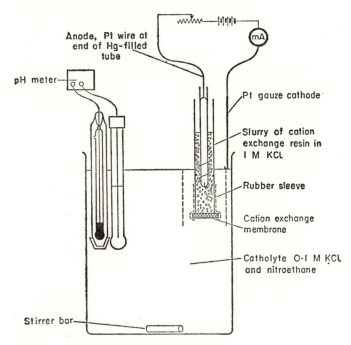

FIG. 1. Cell for amperometric measurement of rapid reaction rate.

Method

A disc of cation exchange membrane is cut out to fit over the end of the glass tube. The hydrogen form of the membrane and resin is converted to the sodium form in the following manner. The resin and membrane are placed in a beaker and a forty-fold excess of a 4 M solution of sodium chloride added (the forty-fold excess is in terms of equivalents of sodium and is with respect to equivalents of hydrogen in the resin and membrane). Sufficient 0·5 M sodium hydroxide is added to make the supernatant liquid neutral to methyl orange and stirred for 5 min. The mixture is neutralized again by adding more sodium hydroxide. This procedure is continued until the indicator colour remains yellow after 5 min stirring. The mixture is decanted and washed until no chloride is detected in the washings.

An aqueous solution is prepared which is accurately 0·1 mole/l. with respect to nitroethane and approximately 0·1 mole/l. with respect to potassium chloride. An accurately measured volume (approximately 250 ml) of this solution is placed in the 400 ml beaker. The electrodes and the platinum gauze cathode are suspended in this solution.

The membrane disc is fixed to one end of the glass tube by means of a rubber sleeve. A slurry of the resin in 1 M potassium chloride is prepared and a sufficient quantity placed in this tube so that the tube is three-quarters full. The platinum wire anode is then suspended within this slurry. The whole assembly is immersed in the catholyte so that it is within the cylindrical platinum gauze cathode (Fig. 1).

A variable voltage of 10–20 V is applied across the anode and cathode and the voltage adjusted so that the milliammeter placed in series with the cell registers a current of 20–80 mA. A steady increase in the pH meter reading will be observed. The voltage is readjusted so that a condition is quickly reached where the current and pH are constant.

A series of values of k may be obtained by varying the voltage slightly and noting the pH meter and milliammeter readings at the steady state.

It will be noted that a minimum current is required to prevent diffusion control of the reaction.

[1] J. A. YOUNG and R. J. ZETO, *J. Chem. Educ.*, **35**, 146–8 (1958).

[2] This cation exchange resin can be obtained from Bio-Rad Laboratories, 32nd and Griffin Avenue, Richmond, California, U.S.A.

[3] This cation exchange membrane can be obtained from Ionics, Inc., 152 Sixth Street, Cambridge 42, Massachusetts, U.S.A.

A Velocity Constant by a Potentiometric Method

Discussion

The reaction between piperidine and 2 : 4-dinitrochlorobenzene takes place in two steps

$$O_2N-\underset{NO_2}{\underset{|}{\bigcirc}}-Cl + H-N\bigcirc \rightarrow O_2N-\underset{NO_2}{\underset{|}{\bigcirc}}-\overset{+}{N}\bigcirc + Cl^-$$
$$\underset{H}{}$$

$$O_2N-\underset{NO_2}{\underset{|}{\bigcirc}}-\underset{\underset{H}{|}}{\overset{+}{N}}\bigcirc + Cl^- + H-N\bigcirc \rightarrow O_2N-\underset{NO_2}{\underset{|}{\bigcirc}}-N\bigcirc + \bigcirc\overset{+}{N}H_2Cl^-$$

The rate equation for the first step which is the slow and hence the rate-determining step is

$$\frac{dx}{dt} = k(a-x)(b-2x) \tag{1}$$

where a and b are the initial concentrations of 2 : 4-dinitrochlorobenzene and piperidine, respectively, and x the number of moles of 2 : 4-dinitrochlorobenzene which have reacted after time t.

If the initial concentration of 2 : 4-dinitrochlorobenzene is half that of the piperidine, equation (1) reduces to

$$\frac{dx}{dt} = 2k(a-x)^2 \tag{2}$$

On integration equation (2) becomes

$$\frac{1}{a-x} = 2kt + C$$

where C is a constant of integration.

The extent of reaction x may be followed by determining the chloride ion concentration after time t by a potentiometric titration.

A graph of $1/(a-x)$ vs. t should then be a straight line having a slope equal to $2k$.

Apparatus and Chemicals

Thermostat at 25°C, three 500 ml graduated flasks, 25 ml pipette, Pye pH meter, burette, six 250 ml beakers, glass electrode, silver–silver chloride electrode, two weighing bottles, 2 : 4-dinitrochlorobenzene, piperidine, 95 per cent ethanol and 0·5 N sulphuric acid.

Method

A 500 ml graduated flask is filled up to the mark with 95 per cent ethanol, placed in a thermostat at 25°C for 15 min and then 25 ml of the ethanol is withdrawn. The requisite quantities of piperidine and 2 : 4-dinitrochlorobenzene are weighed out to give 0·04 M and 0·02 M solutions, respectively, when diluted to 500 ml.

The 2 : 4-dinitrochlorobenzene is added to the ethanol and part of the 25 ml of ethanol is used to wash out any of the adhering liquid. The ethanol–2 : 4-dinitrochlorobenzene solution is allowed to come to thermal equilibrium. The piperidine is then added to this solution, any adhering material being washed out with the remainder of the 25 ml of ethanol. The time of addition of piperidine is noted. The solution is shaken vigorously and made up to the mark with ethanol if necessary.

Approximately a minute from the time of mixing a 25 ml aliquot of the reaction mixture is withdrawn and delivered directly into about 15 ml of 0·5 N sulphuric acid and shaken vigorously. The sulphuric acid stops the reaction immediately. The time is noted at which half the sample has drained into the sulphuric acid. Further 25 ml samples are withdrawn at various times and subjected to the same treatment. The time intervals from the time of mixing should be 1, 5, 10, 20, 40, 60, 80, 100, 120 and 140 min.

The chloride ion concentration in each aliquot is determined by titration with a 0·02 N silver nitrate solution.

A glass electrode and a silver–silver chloride electrode, which are connected to a pH meter, are submerged in the sample. The silver nitrate solution is allowed to run in until there is a sharp change in potential. This point represents the end point for the titration.

In order to obtain accurate end points, three reaction mixtures are prepared and duplicate experiments run simultaneously. Thus having three samples, one can be used to determine an approximate end point whilst the other two can be used for accurate determinations.

A Velocity Constant by a Polarimetric Method

Discussion

The rate expression for the acid catalysed reaction

$$1\text{-menthyl formate} + H_2O \underset{k_2}{\overset{k_1}{\rightleftharpoons}} 1\text{-menthol} + HCOOH$$

is

$$\frac{dx}{dt} = k_1(a-x)(b-x) - k_2 x^2 \tag{1}$$

where a and b are the initial concentrations of 1-menthol and formic acid after time t and k_1 and k_2 the forward and reverse velocity constants, respectively. Now

$$K = \frac{k_1}{k_2} = \frac{x_e^2}{(a-x_e)(b-x_e)} \tag{2}$$

where x_e is the concentration of 1-menthol and formic acid at equilibrium and K the equilibrium constant for the reaction. Substituting for k_2 from equation (2) in equation (1)

$$\frac{dx}{dt} = k_1(a-x)(b-x) - \frac{k_1(a-x_e)(b-x_e)x^2}{x_e^2} \tag{3}$$

On integration, equation (3) becomes

$$t + C = \frac{2 \cdot 303 x_e}{k_1[2ab - (a+b)x_e]} \log \frac{x - x_e/\{[(a+b)/ab]x_e - 1\}}{x - x_e} \tag{4}$$

The constant C is evaluated by applying the condition that $x = 0$ when $t = 0$. Substituting the value of C, so obtained, in equation (4),

$$t = \frac{2 \cdot 303 x_e}{k_1[2ab - (a+b)x_e]} \log\left[\left(\frac{x - x_e/A}{x - x_e}\right)(A)\right] \tag{5}$$

where

$$A = \left(\frac{a+b}{ab}\right)x_e - 1$$

189

The course of this reaction is followed polarimetrically; the degree of rotation α of the reaction mixture being noted at various times t. In order to use the polarimeter readings directly, equation (5) must be modified such that x and x_e in the logarithmic term are expressed in terms of the corresponding polarimeter readings α and α_e.

This leads to the expression

$$t = \frac{2 \cdot 303 x_e}{k_1[2ab - (a+b)x_e]} \log\left(\frac{D-\alpha}{\alpha-\alpha_e}\right) (A)$$

(6)

from which k_1 may be determined by plotting $\log[(D-\alpha)/(\alpha-\alpha_e)]$ vs. t, where

$$D = l[\alpha]_F \frac{w_F}{W} - \frac{l[\alpha]_F(w_F/W) - \alpha_e}{[(a+b)/ab]x_e - 1}$$

$$a = \frac{w_F}{W} \cdot \frac{1000d}{M_F}, \qquad b = \frac{w_W}{W} \cdot \frac{1000d}{M_F}$$

and

$$x_e = \frac{l[\alpha]_F(w_F/W) - \alpha_e}{(l/1000d)\{M_F[\alpha]_F - M_M[\alpha]_M\}}$$

where

$d = $ density of the equilibrium solution (moles/l.),
$l = $ length of the polarimeter tube (dm),
$w_F = $ weight of l-menthyl formate,
$w_M = $ weight of l-menthol,
$w_W = $ weight of water,
$W = $ total weight of solution,
$[\alpha] = $ specific rotation (subscripts as for w),
$M = $ molecular weight (subscripts as for w).

Then k_1 may be determined from the relationship

$$k_1 = \frac{2 \cdot 303 \text{ (slope)}}{2ab - (a+b)x_e}$$

The specific rotations of l-menthyl formate and l-menthol for the particular experimental conditions described are:

$$[\alpha]_F^D = -70 \cdot 0, \qquad [\alpha]_M^D = -43 \cdot 0$$

It is permissible to use absolute values of rotation rather than the true negative values since all calculations involve only ratios of rotation differences.

It is believed that no appreciable error is involved in using the following approximations:

(1) The specific rotation of 1-menthyl formate cannot be measured in a hydrochloric acid solution due to hydrolysis. The value used is that for a solution of 1-menthyl formate in aqueous t.-butyl alcohol.

(2) It is assumed that there is no change in density during the course of the reaction.

FIG. 1. Reaction flask.

Apparatus and Chemicals

Polarimeter, 125 ml reaction flask (Fig. 1), pyknometer, thermostat at 25°C, pure sample of 1-menthyl formate,* concentrated reagent-grade hydrochloric acid, t.-butyl alcohol and 0·1 N sodium hydroxide.

Method

A solution of t.-butyl alcohol and aqueous hydrochloric acid are prepared by adding a weighed amount of hydrochloric acid solution to a weighed amount of t.-butyl alcohol. A ratio of 18 ml hydrochloric acid solution to 9 ml t.-butyl alcohol is used. An aliquot of the t.-butyl alcohol–water–hydrochloric acid mixture is titrated against standard sodium hydroxide solution.

The acid concentration in the final reaction mixture should be approximately 0·37 moles per l.

---

* *Preparation of 1-Menthyl Formate*

1-Menthol (340 g), formic acid (225 g) (98–100 per cent), and toluene (150 ml) are put in a 2-l. round-bottom flask and refluxed for 2 hr on a steam bath. This mixture is then distilled on a steam bath until the rate of distillation becomes slow. At this time the azeotrope of water and toluene comes off at 90°C–95°C. Direct heat is now applied to remove the azeotrope of water and formic acid which boils at about 107°C and the remaining toluene which boils at 110°C. The temperature then rises fairly rapidly to about 219°C. At this point, the residue is transferred to a 500 ml distilling flask and distilled, the distillate boiling between 219°C–220°C (730 mm pressure) being retained. If desired, fractional distillation under reduced pressure may be used. The yield is 86–87 per cent. $D^{25}_{25} = 0.9329$, $n^{D}_{25} = 1.4480$ and $(\alpha)^{D}_{25} = -79.1$.

The polarimeter is prepared for use as described in the instruction pamphlet issued by the manufacturers.

Five millilitres of 1-menthyl formate is run in from a pipette into the centre section of a weighed reaction flask. The stopper is replaced and the flask and contents weighed. Ten millilitres of the t.-butyl alcohol–water–hydrochloric acid mixture is delivered from a pipette into the outer section of the reaction flask. The stopper is replaced and the flask and contents reweighed.

The reaction flask is suspended in a thermostat at 25°C. Whilst awaiting the establishment of thermal equilibrium the zero point for the empty polarimeter tube is recorded. The flask is removed from the thermostat, the contents thoroughly mixed, the time of mixing noted and the reaction mixture poured into the polarimeter tube to a level just below the constriction. A polarimeter reading is taken as soon as convenient and subsequently every 10 or 15 min, the time of each measurement being recorded. When the reaction has slowed down, the time interval is allowed to extend to between $\frac{1}{2}$ hr and 1 hr. Equilibrium is indicated when there is no change in the polarimeter reading over a period of 2 to 3 hr.

The density of the equilibrium solution is determined using a pyknometer (see Experiment 3). The constant k_1 may then be determined.

[1] R. N. SMITH and W. V. BOLLING, *J. Chem. Educ.*, **27**, 369–73 (1950).

A Velocity Constant for an Anionotropic Rearrangement (Spectrophotometric Method)

Discussion

If the absorption properties of the products of a reaction differ significantly from those of the reactants, spectrophotometry offers a ready means of determining the rate of the reaction. Due to the marked increase in the ultra-violet light absorption of systems I and II (where conjugation of the aryl group R can more readily occur with the adjacent double bond than with a single bond)

$$
\begin{array}{cc}
\text{R} & \text{R} \\
\diagdown & \diagdown \\
\text{HO—C—CH=CH}_2 & \text{C=CH—CH}_2\text{OH} \\
\diagup & \diagup \\
\text{H} & \text{H} \\
\text{I} & \text{II}
\end{array}
$$

a reaction involving structures of types I and II as products and reactants can be followed spectrophotometrically.

Phenylallyl alcohol in aqueous dioxan undergoes an anionotropic rearrangement[1] in the presence of dilute mineral acid to form cinnamyl alcohol

$$C_6H_5 . CH(OH) . CH = CH_2 \rightarrow C_6H_5 . CH{=}CH . CH_2OH$$

The change is accompanied by a large increase in the light absorption in the 2500 Å region.

The reaction follows first order kinetics and as the molar absorptivity ϵ is related to the concentration c (mole 1^{-1}.) by the equation

$$\epsilon = \frac{A}{cl} \tag{1}$$

where A is the absorbance and l the thickness of the absorbing solution, the rate constant can be evaluated from the relationship

$$k_1 = \frac{2\cdot303}{t} \log\frac{(\epsilon_\infty - \epsilon_0)}{(\epsilon_\infty - \epsilon_t)} \tag{2}$$

where ϵ_0 is the initial value of the absorptivity, ϵ_∞ the final value, and ϵ_t the value after a time t min.

Apparatus and Chemicals

Spectrophotometer e.g. Unicam SP.500, 250 ml graduated flask, 50 ml graduated flask, 1·0 M hydrochloric acid, 0·5 M sodium acetate, phenylallyl alcohol,* purified dioxan (by refluxing over sodium, fractionating, and storing in the dark under nitrogen).

Method

A small quantity (25 mg) of phenylallyl alcohol is weighed out accurately into a 50 ml flask. Purified dioxan is added to the flask and the solution adjusted to 50 ml. A further quantity of dioxan (74 ml) is added to the 250 ml graduated flask. Subsequently, 12·8 ml of 1·0 M hydrochloric acid and 38·3 ml of distilled water are added to the 250 ml flask. The flask is shaken and placed in a thermostat at 40°C for 1 hr.

A measured volume (2·5 ml) of the dioxan solution of the phenylallyl alcohol is added to the acidified dioxan solution. The flask is shaken vigorously in the thermostat. A 5 ml portion is withdrawn with a pipette. The reaction is stopped by mixing this portion of the solution with 1 ml of sodium acetate solution. The time is noted.

The absorbance of this solution is determined at 2510 Å.

This operation is repeated a number of times at intervals of 45 min. The final operation is made 24 hr after the commencement of the reaction.

The molar absorptivities of the solutions are calculated from equation (1). The rate constant is evaluated using equation (2).

* Phenylallyl alcohol may be prepared by the slow addition of phenyl magnesium bromide in ether to an ethereal solution of acraldehyde at 0°C, hydrolysis with cold saturated aqueous ammonium chloride solution, removal of the solvent, and purification by distillation, etc.

b.p. 53−54 °C/0·15 mm : $n_D^{14·50} = 1·5464$.

[1] E. A. BRAUDE, E. R. H. JONES and E. S. STERN, *J. Chem. Soc.*, 396 (1946).

Gas Adsorption (McBain–Bakr Balance)

Discussion

Adsorption from the liquid phase has already been described in Experiment 36, where the empirical Freundlich adsorption isotherm is used. Neither the Freundlich nor Langmuir expressions, however, are adequate when investigating the results of adsorption measurements of vapours on porous solids at relatively high vapour pressures. More complicated expressions are used, one of the best known being that due to Brunaer, Emmett and Teller[1] (the B.E.T. isotherm). This equation was deduced making the assumption that multi-molecular layers could form on the surface of the adsorbent. The equation can be expressed in the form

$$\frac{p/p_0}{x[1 - (p/p_0)]} = \frac{1}{x_m c} + \frac{c-1}{x_m c} \cdot \frac{p}{p_0} \tag{1}$$

where x is the weight of vapour adsorbed per gram of adsorbent at a partial pressure p, p_0 the saturated vapour pressure of the adsorbate at the temperature of the experiment, x_m the weight of vapour in grams which would be required to form a monomolecular layer over the surface, and c a constant related to the heat of condensation. This B.E.T. isotherm represents fairly satisfactorily the adsorption of many vapours on finely divided powders.

The amount of vapour adsorbed at various pressures may be measured with a McBain–Bakr[2] balance by noting the extension of a quartz spring due to the increase in weight of the powder on which adsorption takes place. The powder is contained in a glass bucket which is attached to the end of the spring, see Fig. 1.

As seen from equation (1) a plot of $(p/p_0)/x(1-p/p_0)$ vs. p/p_0 should give a straight line of slope $(c-1)/x_m c$ and intercept $1/x_m c$. A linear plot is usually obtained for p/p_0 values in the range 0·05 to 0·35. The value of x_m may be obtained from these data and hence the surface area S of the powder calculated from equation (2)

$$S = \frac{aN x_m}{M} \mathring{A}^2 \, g^{-1} \tag{2}$$

where a is the molecular cross-sectional area of the adsorbate, N Avogadro's number and M the molecular weight of the adsorbate.

Apparatus and Chemicals

Glass apparatus as shown in Fig. 1, with rotary and diffusion pumps, manometer and McLeod or Pirani gauge attachments, cathetometer, quartz spring of sensitivity 20 cm per g load, Tesla coil, silica gel and benzene.

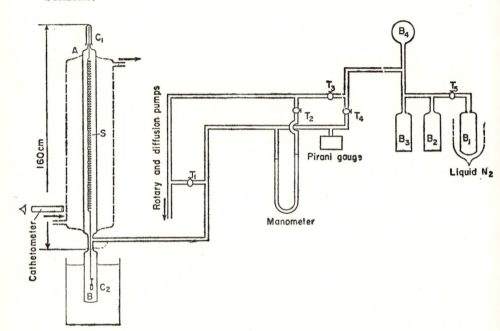

FIG. 1. Adsorption apparatus (the McBain–Bakr balance).

Method

The quartz spring is attached to the glass hook A by removing the cover C_1. After the tube C_2 is removed the small bucket B is attached to the other end of the spring by means of a nylon fibre. This spring S and the bucket B are enclosed by suitable thermostats. The pumping system and the glass assembly are arranged so that either side of the system, i.e. the column containing the spring, the manometer and the McLeod gauge, or the distillation assembly containing B_1, B_2, B_3 and B_4 can be isolated.

When the apparatus is finally assembled:

(a) The spring is calibrated by measuring the extension for known added weights in the bucket B.

(b) All the vacuum taps are opened and the system pumped out; a vacuum of 10^{-4} mm Hg must ultimately be obtained. The rotary pump is first used to obtain a vacuum of 10^{-2} mm and is then coupled with the diffusion pump to achieve a final pressure of 10^{-4} mm Hg or lower. Before the diffusion pump is used the apparatus can be tested for leaks using the Tesla coil.

(c) For the vacuum distillation the taps T_1, T_2, T_3, T_4 and T_5 are closed. The removable bulb B_1 is then filled with benzene and re-attached to the system. The flask containing the liquid nitrogen is then placed around this bulb B_1. Taps T_5 and T_3 are then opened and the pumps are switched on. When a good vacuum has been obtained tap T_3 is closed. The flask containing the liquid nitrogen is then placed around the bulb B_2 or B_3. When all the liquid has distilled from B_1, the tap T_5 is closed and tap T_3 opened in order to remove any air which may have been dissolved in the benzene. This air is then removed by the pumping system. The tap T_3 is closed and the reservoir B_4 becomes full of vapour when the liquid nitrogen is removed.

(d) After admitting air the flask C_2 is removed and 0·25 g of silica gel is placed in the glass bucket B. The column containing the spring is then pumped out with the taps T_2, T_3, T_4 and T_5 closed and the tap T_1 is opened. When the final pressure of 10^{-4} mm Hg is obtained, the extension of the spring is noted and the new weight of the adsorbate, due to the de-gassing, calculated from the calibration.

Tap T_1 is closed and benzene vapour is admitted to the column by opening tap T_4. Small volumes of the vapour are then consecutively admitted to the column and the degree of adsorption measured by noting the spring extension. The actual weight adsorbed x can be obtained from the original weight calibration. The vapour pressure p is recorded on the manometer for each adsorption, p_0 the saturation pressure of benzene is known at the temperature of the experiment.[3]

A graph of $(p/p_0)/x(1-p/p_0)$ vs. p/p_0 is plotted and the slope and intercept of the resultant straight line determined. Therefore the value of x_m can be calculated. The surface area S of the silica gel can also be calculated from equation (2) since the molecular cross-sectional area of benzene at 25°C is 32 Å^2.

Note: Benzene adsorption[4] may not give the correct surface area values because of possible interaction between the adsorbent and adsorbate, i.e. possible polarization of the benzene. The most satisfactory results are given by nitrogen adsorption at liquid nitrogen temperatures.

[1] S. BRUNAER, P. H. EMMETT and E. TELLER, *J. Amer. Chem. Soc.*, **60**, 309 (1938).

[2] J. W. McBAIN and A. M. BAKR, *J. Amer. Chem. Soc.*, **48**, 690 (1926).

[3] C. D. HODGMAN (Ed.), *Handbook of Chemistry and Physics*, Chemical Rubber Publishing Co.

[4] D. M. HORVAT and K. S. W. SING, *J. Appl. Chem.* **11**, 313 (1961).

Surface Tension–Concentration Relationship for Solutions (Gibbs Equation)

Discussion

The positive adsorption of a solute by a suitable adsorbent decreases the surface (or interfacial) tension; negatively adsorbed solutes (e.g. salts) causing an increase. For the special case of a very dilute solution of a non-ionic surface-active substance, the quantitative relationship between concentration, adsorption, and change of surface tension is given by the Gibbs adsorption equation:

$$\Gamma = \frac{-a}{RT}\frac{d\gamma}{da} \backsimeq \frac{-c}{RT}\frac{d\gamma}{dc} \tag{1}$$

where Γ is the mass of solute adsorbed per unit area of surface from a solution of concentration c (activity a) and surface tension γ. Hence if the surface tension of a liquid is measured at constant temperature for various concentrations $d\gamma/dc$ may be obtained for any concentration from the tangents to the graph of γ vs. c. The various Γ values may then be calculated from equation (1).

The value of Γ tends to a limit at the higher concentrations and this limiting value Γ_{Lt} can be estimated. The reciprocal value $1/\Gamma_{Lt}$ can be calculated, hence the area A, in the surface per molecule of adsorbed solute may be calculated from

$$A = \frac{1}{N\Gamma_{Lt}} \tag{2}$$

where N is Avogadro's number.

The surface tension may be measured with a du Nouy tensiometer with which the pull on a platinum ring due to surface tension is measured with a torsion balance. The detachment force is related to the surface (or interfacial) tension by the expression

$$\gamma = \frac{\beta f}{4\pi r} \tag{3}$$

where f is the pull (in dynes) on the ring, r is the mean radius of the ring and β is a correction factor. The factor β allows for the non-vertical direction of the tension forces and for the complex shape of the liquid supported by the ring at the point of detachment. It thus depends on the dimensions of the ring and the nature of the interface. For simplicity in this experiment β may be assumed to have the value unity.

Apparatus and Chemicals

Du Nouy tensiometer, amyl alcohol.

Method

A $0 \cdot 1$ M solution of n-amyl alcohol is made containing $1 \cdot 08$ ml of alcohol per 100 ml of solution and from this solution $0 \cdot 08$ M, $0 \cdot 04$ M, $0 \cdot 02$ M and $0 \cdot 01$ M solutions are prepared by dilution. The surface tensions of pure n-amyl alcohol, water and the solutions are measured. The temperature is recorded.

A graph is then plotted of surface tension γ vs. concentration c, and a smooth curve drawn through these points. Tangents are then constructed at concentrations $0 \cdot 01$, $0 \cdot 02$, $0 \cdot 04$, $0 \cdot 06$ and $0 \cdot 08$ M, and the values of $d\gamma/dc$ determined at these points. Values of Γ are then calculated from equation (1) and a graph of these Γ values plotted against the appropriate concentrations of the amyl alcohol solutions. The limiting value Γ_{Lt} can then be estimated and the reciprocal $1/\Gamma_{Lt}$ calculated.

The value of $1/\Gamma$ is calculated for each experimental concentration and hence the area per molecule A is found. A graph of Π against A is plotted where Π is the surface pressure (γ_{H_2O}—$\gamma_{solution}$). If the results fit the equation

$$\Pi(A - A_0) = kT$$

the value of A_0 should be calculated.

Molecular Weight of a Polymer from Viscosity Measurements

Discussion

The ratio of the viscosity η of a solution of non-spherical high polymer molecules to the viscosity η_0 of the solvent is related to the molecular weight M of the polymer by the expression

$$\frac{(\eta/\eta_0) - 1}{c} = KM^\alpha \tag{1}$$

In this equation K is a constant for any given type of polymer, solvent and temperature, α is a function of the geometry of the molecule, and c is the number of grams of polymer in 100 ml of solution. The term $(\eta/\eta_0) - 1$ is known as the specific viscosity η_{sp}; equation (1) may therefore be written in the form

$$\frac{\eta_{sp}}{c} = KM^\alpha \tag{2}$$

This equation is only valid for very dilute solutions (less than 1 per cent) and hence the graph which is drawn of η_{sp}/c vs. c is extrapolated to zero concentration. The extrapolated value is known as the intrinsic velocity $[\eta]$, i.e.

$$[\eta] = \lim_{c \to 0} . \frac{\eta_{sp}}{c} \tag{3}$$

If the logarithmic function $\ln(\eta/\eta_0)$ is expanded as an infinite series, since the second and higher terms can be neglected as the concentration approaches zero, it will be seen that

$$\lim_{c \to 0} . \frac{\eta_{sp}}{c} = \lim_{c \to 0} . \frac{1}{c} \ln\frac{\eta}{\eta_0} \tag{4}$$

hence

$$[\eta] = \lim_{c \to 0} . \frac{1}{c} \ln\frac{\eta}{\eta_0} \tag{5}$$

The intrinsic viscosity is therefore the intercept on the graph of either η_{sp}/c or $(1/c) \ln(\eta/\eta_0)$ vs. c. A more reliable value of the intercept is obtained by drawing both graphs.

The $[\eta]$ value obtained by this method may then be used to determine the molecular weight of the polymer by applying the equation

$$[\eta] = KM^\alpha \tag{6}$$

Apparatus and Chemicals

Ostwald viscometer, stop watch, thermostat at 25°C, polystyrene and toluene.

Method

The technique of measuring viscosity with an Ostwald viscometer is discussed in Experiment 4. For this experiment, equation (3) in Experiment 4 takes the form

$$\frac{\eta}{\eta_0} = \frac{dt}{d_0 t_0} \tag{7}$$

where t and t_0 are the times of flow of the solution and solvent respectively, and d and d_0 are the corresponding densities. At these dilutions, however, the densities of solution and pure solvent can be considered to be the same and equation (7) reduces to

$$\frac{\eta}{\eta_0} = \frac{t}{t_0}$$

For this particular experiment the flow time of pure toluene is first determined.

A solution containing 500 mg of polystyrene in 25 ml of toluene (2·0 g/100 ml) is prepared and the flow time of this solution is determined. Further solutions are prepared by dilution of the original solution. Concentrations of 1·6, 1·2, 0·8 and 0·4 g of polystyrene per 100 ml of toluene are recommended.

The details regarding the cleaning of the viscometer and the thermal equilibration of the solutions are outlined in Experiment 4.

The results are recorded in tabular form under the headings:

c	t	$\dfrac{\eta}{\eta_0} = \dfrac{t}{t_0}$	$\dfrac{\eta}{\eta_0} - 1 = \eta_{sp}$	$\dfrac{\eta_{sp}}{c}$	$\dfrac{1}{c} \ln \dfrac{\eta}{\eta_0}$

For solutions of polystyrene in toluene at 25°C the constants in equation (6) have the following values:

$$K = 3\cdot7 \times 10^{-4}$$
$$\alpha = 0\cdot62$$

If scattered results are obtained, the solutions of the polymer and the solvent should be centrifuged.

Electrophoresis

Discussion

If a background electrolyte is allowed to flow down a sheet of filter paper across which an electric field is applied, then an ion applied at an intermediate position on the paper will deviate laterally in accordance with its inherent mobility. This principle is known as electrophoresis and can be applied to the separation of a mixture of ions.

An apparatus which may be used to demonstrate this principle is represented diagrammatically in Fig. 1. The filter paper sheet A is suspended from a trough B in which a constant level of electrolyte is maintained. The test substance is in a small container C from which it is fed to the filter paper sheet by means of a paper wick D.

Electrically neutral species will follow a straight path in the direction of the electrolyte flow whilst those which are positively charged will deviate towards F and those negatively charged towards E. The extent of deviation will be governed by the size and charge of the various species.

Apparatus and Chemicals

Shandon Continuous Electrophoresis Apparatus, Power Pack and Dipping Tray, electric fan, drying oven, Whatman 3 M.M. paper,* 10 l. of 1 M acetic acid, pyridine, bromine, 0·2 per cent solution of ninhydrin in acetone, 0·5 M sodium hydroxide, 0·2 per cent solution of isatin in acetone, 0·1 per cent solution of 8-hydroxy quinoline in acetone and 50 ml of an aqueous mixture of 0·04 M arginine, 0·04 M glycine and 0·04 M aspartic acid.

Method

The continuous electrophoresis apparatus is set up as described in the instruction pamphlet. The reservoirs, trough B and electrode vessels G and H are filled with 1 M acetic acid. The paper sheet is folded 1·5 cm from the non-serrated end and then folded again in the opposite direction 8 cm from the same end. The filter paper sheet is placed so that the first fold dips into trough B where it is held in position by a

* This paper is supplied by Reeve Angel and Co. Ltd.

heavy glass rod. The second fold rests on the glass rod at the edge of
the trough.

The cabinet is completely closed and the paper allowed to become
saturated with acetic acid solution. During this period of saturation
the working potential difference of 1000 V is applied across the paper.
This will produce a current of about 5–6 mA once the filter paper is
saturated with buffer.

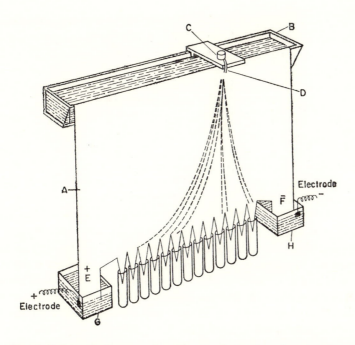

FIG. 1. Schematic diagram of continuous electrophoresis apparatus.

The specimen container is prepared for use by cutting a piece of
filter paper as shown in Fig. 2, wrapping it around a short length of
wick and inserting this into the spout. The container is then filled
with the solution of amino acids. The power pack is switched off, the
cabinet is opened and the container placed in the holding bracket so
that the wick touches the paper about 3 cm from the anode side.
The cabinet is closed and electrophoresis is allowed to proceed for
18 hr.

At the end of this time, the power is switched off, the paper removed
from the cabinet and the acetic acid solution evaporated off. The
evaporation is carried out in a fume cupboard using an electric fan.

The dried paper is dipped in a 0·2 per cent solution of ninhydrin in acetone to which a few drops of pyridine have been added and then hung in an oven at about 100°C for 5 min. Colour will develop in curved bands indicating the paths along which the individual amino acids have migrated.

In order to determine the extent of separation and to identify the amino acid corresponding to each path, the following tests are applied to the solutions in the appropriate test tubes.

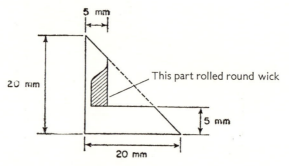

FIG. 2. The paper wick.

A spot of solution is placed on a filter paper strip and dipped into a 0·1 per cent solution of 8-hydroxy quinoline in acetone. The acetone is allowed to evaporate. The paper is then dipped into a solution of 0·3 ml of bromine in 100 ml of 0·5 M sodium hydroxide and allowed to dry. An orange-red colour indicates the presence of arginine. This is an application of the Sakaguchi reaction which is exclusive for mono-substituted guanidines.

Again a spot of solution is placed on a strip of filter paper and the paper dipped in a 0·2 per cent solution of isatin in acetone to which a few drops of pyridine have been added. The paper is heated at 105°C for 2–3 min. A blue colour indicates the presence of aspartic acid. This reaction is given by a number of other amino acids but not by glycine or arginine.

The coloured band leading to the samples which give neither of the above reactions corresponds to the path followed by the glycine molecules.

Variation of Conductance with Concentration

Discussion

For the variation of conductance with concentration the Onsager equation is

$$\Lambda = \Lambda_0 - (B_1\Lambda_0 + B_2)\sqrt{c} \tag{1}$$

giving

$$\Lambda_0 = \frac{\Lambda + B_2\sqrt{c}}{1 - B_1\sqrt{c}} \tag{2}$$

where B_1 and B_2 are constants depending upon the nature of the solvent, the temperature and I the ionic strength. For a uni-univalent electrolyte in water at 25°C

$$B_1 = \frac{8 \cdot 20 \times 10^5}{(\epsilon T)^{3/2}} = 0 \cdot 230 \qquad B_2 = \frac{82 \cdot 4}{\eta(\epsilon T)^{1/2}} = 60 \cdot 65$$

$$I = \tfrac{1}{2}\{c(1)^2 + c(1)^2\} = c$$

where T is the absolute temperature, ϵ the dielectric constant and η the viscosity of water.

These equations are satisfactory for uni-univalent electrolytes up to concentrations of 0·001 N. (See Experiment 37.)

For more concentrated solutions Robinson and Stokes[1] have suggested a modification of equation (1). As ions are not merely point charges but have a finite size, Robinson and Stokes have divided the coefficient $(B_1\Lambda_0 + B_2)$ by the quantity $(1 + \kappa a)$ giving the modified equation.

$$\Lambda = \Lambda_0 - \frac{B_1\Lambda_0 + B_2}{1 + \kappa a}\sqrt{c} \tag{3}$$

where a is an ion size parameter and κ, not to be confused with specific conductance, is a fundamental quantity of the Debye–Hückel theory. (The quantity κ is a function of concentration, ionic charge, temperature and the dielectric constant of the solvent and has the dimensions of reciprocal length.) κ is proportional to the square root of the ionic strength, and for uni-univalent electrolytes is given by

$$\kappa = B\sqrt{c} \tag{4}$$

where

$$B = \frac{50 \cdot 29 \times 10^8}{(\epsilon T)^{1/2}}$$

and c is the concentration in moles per litre.

The Robinson and Stokes equation can be re-written

$$\Lambda_0 = \Lambda + \frac{B_1\Lambda + B_2}{1 + (aB - B_1)\sqrt{c}}\sqrt{c} \tag{5}$$

This equation can be used to determine the value of Λ_0.

For aqueous solutions of uni-univalent electrolytes equations (3) and (5) are satisfactory for concentrated solutions up to $0 \cdot 1$ N.

Apparatus and Chemicals

Wayne Kerr Universal Bridge,[2] conductivity cell, thermostat, $0 \cdot 01$ D potassium chloride, $0 \cdot 1$ N sodium chloride.

Method

The initial zero balance for the bridge is determined as stated in the Wayne Kerr Instruction Manual.

The cell containing $0 \cdot 01$ D potassium chloride is placed in a thermostat at 25°C and is connected to the bridge. The cell constant is determined as in Experiment 37.

The experiment is repeated with $0 \cdot 1$ N sodium chloride using the same amount of solution in the conductivity cell. The solution is diluted to give solutions of concentration $0 \cdot 075$, $0 \cdot 05$ and $0 \cdot 01$ N and the conductances of the solutions are determined.

The equivalent conductances Λ_1, Λ_2, Λ_3 and Λ_4 are calculated for these four solutions of sodium chloride.

The value of the equivalent conductances Λ_1, Λ_2, Λ_3 and Λ_4 are calculated from the Onsager equation (1) where

$$\Lambda_0 = 126 \cdot 45 \text{ mho cm}^{-1} \text{ eqt.}^{-1}, \qquad B_1 = 0 \cdot 230, \qquad B_2 = 60 \cdot 65$$

The value of the equivalent conductances Λ_1, Λ_2, Λ_3 and Λ_4 are calculated from the Robinson and Stokes equation (3) taking

$$\kappa = B\sqrt{c}, \qquad 10^{-8}B = 0 \cdot 3291, \qquad a = 4\text{Å}$$

The experimental conductances are compared with those calculated from the theoretical equations.

[1] R. A. ROBINSON and R. H. STOKES, *J. Amer. Chem. Soc.*, **76**, 1991 (1954); and R. A. ROBINSON and R. H. STOKES, *Electrolyte Solutions*, 2nd ed., p. 468, Butterworths Scientific Publications.
[2] R. CALVERT, J. A. CORNELIUS, V. S. GRIFFITHS and J. A. STOCK, *J. Phys. Chem.*, **62**, 47 (1958).

Transport Numbers
(Moving Boundary Method)

Discussion

If the number of equivalents of an ion which are transferred during electrolysis are known in terms of the total amount of electricity passed it is possible to calculate the transport number of that ion. In the following method the migration of an ion under the influence of an electric field is measured by observing the movement of a boundary between the solution containing the ion and an indicator solution.

Suppose the boundary moves through a volume of solution v ml in a time t sec and that during this time the current is kept constant at a value of I A. Then, It coulombs or It/F faradays will flow past every point in the circuit in the time t.

The fraction of this quantity of electricity carried by the cation will be equal to its transport number n_c. Hence Itn_c/F equivalents of cation will pass every point in the apparatus. If the concentration of the solution is c eqt./ml then the cation will sweep through a volume of solution Itn_c/F ml in the time t. Hence

$$v = \frac{Itn_c}{cF}$$

or

$$n_c = \frac{cvF}{It} \tag{1}$$

Apparatus and Chemicals

Moving boundary cell, milliammeter, stop clock, d.c. supply (0–200 V), 0·1 N hydrochloric acid and methyl orange.

The cell is illustrated in Fig. 1. It consists of a graduated capillary tube (e.g. 1 ml pipette) containing the hydrochloric acid solution which connects a copper anode with a silver–silver chloride cathode. The cathode is located in an electrode vessel which is connected with the capillary tube so that the denser solution formed around the silver–silver chloride electrode does not diffuse into the capillary. For more accurate work the graduated tube may be immersed in water so that

207

the heat generated during electrolysis may be dissipated, thus reducing the variation in the temperature.

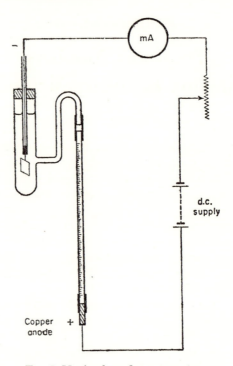

Fig. 1. Moving boundary apparatus.

Method

The graduated tube is cleaned and thoroughly rinsed with distilled water. Sufficient methyl orange is added to a portion of the 0·1 N hydrochloric acid solution until the colour is dense enough to be observed in the thickness of solution which will be contained in the capillary tube. The graduated tube is rinsed out with this hydrochloric acid solution and then attached to the copper anode assembly. The tube is now filled with the hydrochloric acid solution ensuring that no air bubbles adhere to the surface of the anode or the walls of the tube. The electrode vessel which will contain the silver–silver chloride electrode is attached to the upper end of the graduated tube and more hydrochloric acid is added to fill the vessel. Finally the silver–silver chloride electrode is fitted.

A potential is applied to the cell such that a current of about 5 mA flows through the apparatus. The anode will dissolve and the boundary between the copper chloride solution so formed and the hydrochloric acid will move up the tube due to the migration of the cations. The boundary may be easily observed by the difference in colour of the two solutions. As the experiment proceeds the resistance of the electrolyte will increase and the applied potential will have to be continually increased in order to maintain a constant current through the apparatus.

The times taken for the boundary to pass successive graduation marks on the tube should be noted and the average of these time intervals should be used in the calculation.

From the observed data the transport numbers of the hydrogen and chloride ions in 0·1 N hydrochloric acid at the temperature of the experiment are calculated using equation (1).

If a more accurate measure of the current is required than that provided by the milliammeter a standard fixed resistance of value about 300 ohms should be included in the circuit. The current may then be calculated from the potential drop across the standard resistance, which may be measured with a potentiometer.

Transport Numbers (e.m.f. Method)

Discussion

In the concentration cell

$$\text{Zn/ZnSO}_4(a_{\pm})_1 \ \vdots \ \text{ZnSO}_4(a_{\pm})_2/\text{Zn}$$

the dotted line represents the junction between zinc sulphate solution of mean ionic activity $(a_{\pm})_1$ and zinc sulphate solution of mean ionic activity $(a_{\pm})_2$. Across this boundary diffusion of the ions will occur from the more concentrated to the more dilute solution. If the zinc and sulphate ions do not diffuse at the same rate, separation of charge will occur and a difference of potential will be established across the boundary. This is known as a liquid junction potential and will be included in the observed value of the cell e.m.f., E_a, which is given by the relation

$$E_a = t_- \frac{RT}{F} \ln \frac{(a_{\pm})_2}{(a_{\pm})_1} \tag{1}$$

where t_- is the average transport number of the sulphate ion between the two concentrations at which it exists in each half cell.

A concentration cell without a liquid junction may be constructed. In such a cell there is no contact between electrolytes of different concentrations and hence no possibility of diffusion leading to the creation of a liquid junction potential. This type of cell is exemplified by a Helmholtz double cell.

$$\text{Zn/ZnSO}_4(a_{\pm})_1, \quad \text{Hg}_2\text{SO}_4(s)/\text{Hg}/\text{Hg}_2\text{SO}_4(s), \quad \text{ZnSO}_4(a_{\pm})_2/\text{Zn}$$

The double cell may be considered as two simple zinc–mercury cells (or Clark cells) connected in opposition. This is apparent from Fig. 1 which shows how the double cell may be constructed in practice.

The e.m.f., E_b, of the double cell is given by the expression

$$E_b = \frac{RT}{F} \ln \frac{(a_{\pm})_2}{(a_{\pm})_1} \tag{2}$$

Dividing equation (1) by equation (2)

$$\frac{E_a}{E_b} = t_- \tag{3}$$

Thus, from a measurement of the e.m.f.'s of the two concentration cells, the transport number of the sulphate ion in zinc sulphate may be determined.

Apparatus and Chemicals

Tinsley potentiometer, thermostat, zinc electrodes, mercury, mercurous sulphate, 0·5 M and 0·1 M solutions of zinc sulphate saturated with mercurous sulphate.

Method

In order to obtain reproducible results the Clark cells should be freshly prepared for each experiment. Two or three hours will be required for each cell to reach equilibrium and give reproducible values of e.m.f.

The two Clark cells are set up as shown in Fig. 1, the mercury electrodes of each cell being connected together with a piece of wire. The assembly is then placed in a thermostat at 25°C and the e.m.f. of the double cell is measured from time to time until consistent results are obtained.

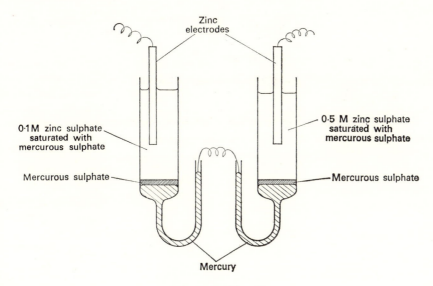

FIG. 1. Double cell.

The double cell is converted to the simple concentration cell by removing the connection between the mercury electrodes and joining the two solutions with a tube containing 0·5 M zinc sulphate solution.

H

This tube should contain a cotton wool plug at the end which is immersed in the 0·1 M zinc sulphate solution to prevent excessive diffusion. This conversion is illustrated in Fig. 2. The e.m.f. of this cell is now measured at intervals until once again consistent results are obtained.

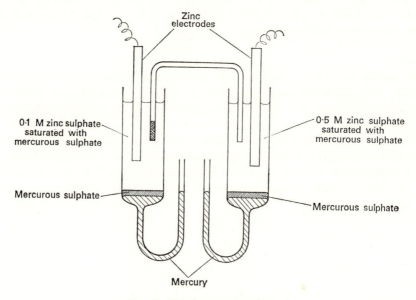

FIG. 2. Simple concentration cell.

From the results, the average transport number of the sulphate ion in zinc sulphate at 25°C is calculated using equation (3).

The Absorption Curve of an Indicator as a Function of pH

Discussion

The study of the absorption curves of an indicator at different pH values provides an excellent method for studying the colour changes occurring in acid–alkali indicators. If the absorption curves (A or ϵ vs. λ), are plotted for an indicator such as bromophenol blue, at a series of pH values, two maxima are obtained on each curve. Moreover, the several curves cross very nearly at a common point called the isosbestic point. The existence of an isosbestic point indicates that a chemical equilibrium exists between two species which are interconvertible.

An indicator is usually a weak organic acid or base which exists in two tautomeric forms. Only one of these tautomers ionizes. Considering an indicator which is a weak acid, these equilibria may be represented:

$$HA_1 \rightleftharpoons \underbrace{HA_2 \rightleftharpoons H^+ + A_2^-}$$

$$\text{Colour } A \qquad \text{Colour } B$$

where the subscripts 1 and 2 represent the two tautomeric forms of the organic part of the molecule.

The equilibrium constant for the first equilibrium K_1 is given by:

$$K_1 = \frac{(HA_2)}{(HA_1)}$$

and the equilibrium constant for the ionization K_2 is given by

$$K_2 = \frac{(H^+)(A_2^-)}{(HA_2)}$$

The product of these two equilibrium constants gives another constant K known as the indicator constant

$$K_1 K_2 = K = \frac{(H^+)(A_2^-)}{(HA_1)}$$

One of the conditions for a substance to be a good indicator is that the concentration of unionized HA_2 should be small. If this is so, we may take it that all of colour B (the alkaline colour) is due to the A_2^- ions.

Assuming that the fraction of indicator in the form of unionized HA_2 is negligible and putting the fraction of indicator in the form of A_2^- ions equal to x then

$$K = \frac{x}{1-x} H^+$$

or

$$\log\frac{x}{1-x} = pH + \log K \tag{1}$$

At a given wavelength, preferably near a maximum and not near an isosbestic point, if A_1 is the absorbance in the most acid solution, A_2 the absorbance in the most alkaline solution and A is the absorbance in a solution of intermediate pH then

$$A = (1-x)A_1 + xA_2$$

Hence

$$x = \frac{A_1 - A}{A_1 - A_2} \tag{2}$$

Substituting in equation (1)

$$\log\frac{A_1 - A}{A - A_2} = pH + \log K \tag{3}$$

A plot of the L.H.S. against pH will be linear and $\log K$ may be determined from the intercept.

Apparatus and Chemicals

Spectrophotometer, e.g. Unicam SP.600, Cambridge pH meter, 1 M hydrochloric acid, 1 M sodium acetate solution, bromophenol blue stock solution (0·6 g indicator, 15 ml 0·1 M sodium hydroxide made up to 1 l. with distilled water).

Method

A quantity of the sodium acetate solution (50 ml) is placed in a 250 ml graduated flask and the solution diluted with distilled water to about 200 ml. A small amount of the stock solution of the indicator (2 ml) is added to the solution and the volume adjusted to 250 ml.

The absorbance of the solution in the range 650–350 mμ at intervals of 10 mμ is determined.

The pH of the solution is determined.

A plot of absorbance A against wavelength λ at this pH value is drawn.

A small amount of 1 M hydrochloric acid is added to reduce the pH of the solution. The absorbance of this solution is determined within the above range of wavelengths. The pH of this solution is measured.

This procedure is repeated until a total of five absorption curves has been obtained over the pH range 2·5–5·5.

The isosbestic point and the regions of maximum absorption are noted. At a wavelength corresponding to a maximum a plot of $\log\left[(A_1 - A)/(A - A_2)\right]$ against pH is drawn. The graph should be linear and at pH $= 0$ the value of the L.H.S. of equation (3) is equal to $\log K$.

Dissociation Constant of an Acid
(Accurate e.m.f. Method)

Discussion

An accurate method for determining the dissociation constant of an acid is available from e.m.f. measurements.[1] Suppose it is required to determine the dissociation constant of a weak acid HA. The following cell may be assembled

$$\text{Pt; quinhydrone|HA}(M), \text{ NaCl}(m), \text{ AgCl|Ag} \qquad \text{(A)}$$

where M and m are the number of moles of acid and sodium chloride, respectively, in 1000 g of water. Let the e.m.f. of this cell be E_A. The above cell is used in conjunction with the simpler cell

$$\text{Pt; quinhydrone|HCl(0·01 M), AgCl|Ag} \qquad \text{(B)}$$

the e.m.f. of which is E_B, measured under identical experimental conditions to those used in the determination of E_A.

If the cells A and B were combined to form the double cell

$$\text{Pt; quinhydrone|HA}(M), \text{ NaCl}(m), \text{ AgCl|Ag|AgCl,}$$
$$\text{HCl(0·01 M)|quinhydrone; Pt.}$$

the e.m.f. of this cell would be $E_A - E_B$. Neglecting the differences in the activities of quinone and hydroquinone in the first cell, and the activities of these two substances in the second cell, we may write

$$\frac{F(E_A - E_B)}{2\cdot303RT} = \log\frac{(0\cdot01)^2(\gamma_{\text{HCl},0\cdot01})^2}{m_{\text{H}^+} \cdot m \cdot (\gamma_{\text{HCl}})^2} \qquad (1)$$

where $\gamma_{\text{HCl},0\cdot01}$ is the mean ionic activity coefficient of 0·01 M hydrochloric acid, m_{H^+} is the molality of the hydrogen ions in cell B and γ_{HCl} is the mean ionic activity coefficient of hydrochloric acid of molality m_{H^+} in the presence of sodium chloride and undissociated HA molecules of molality $(M - m_{\text{H}^+})$. Values of $\gamma_{\text{HCl},0\cdot01}$ can be obtained from the literature[2] but values of γ_{HCl} are not easily determined. This latter difficulty is overcome by using a limiting value of γ_{HCl} as M

tends to zero. This limiting value of γ_{HCl} will be the activity coefficient of hydrochloric acid in a solution of pure sodium chloride. It is written γ^0_{HCl} and values are readily available.[3]

Using the quantity γ^0_{HCl} we may re-write equation (1)

$$\frac{F(E_A - E_B)}{2 \cdot 303 RT} = \log \frac{(0 \cdot 01)^2 (\gamma_{HCl, 0 \cdot 01})^2}{m'_{H^+} \cdot m \cdot (\gamma^0_{HCl})^2} \tag{2}$$

where m'_{H^+} is called the apparent hydrogen ion concentration and is defined by the relationship

$$m'_{H^+} \cdot (\gamma^0_{HCl})^2 = m_{H^+} \cdot (\gamma_{HCl})^2 \tag{3}$$

From measurements of E_A and E_B it is thus possible to calculate m'_{H^+} from equation (2).

The thermodynamic dissociation constant K of the weak monobasic acid HA is defined by

$$K = \frac{\gamma_{H^+} \gamma_{A^-} m_{H^+} m_{A^-}}{\gamma_{HA} m_{HA}} \tag{4}$$

Putting $\gamma_{H^+} \gamma_{A^-} = \gamma^2_{\pm}$, taking $m_{H^+} = m_{A^-}$ and as the ionic strength of the medium is not too high, the activity coefficient of the undissociated HA may be taken as unity.

$$K = \gamma^2_{\pm} \frac{m^2_{H^+}}{M - m_{H^+}} = \gamma^2_{\pm} \cdot k \tag{5}$$

At infinite dilution $\gamma_{\pm}$ will be equal to unity and hence as the ionic strength of the solution decreases k will approach K.

The logarithmic form of equation (5) is

$$\log K = 2 \log \gamma_{\pm} + \log k \tag{6}$$

Further, the activity coefficient of any uni-univalent electrolyte referred to unity at infinite dilution is related to the ionic strength, I, of the solution by the semiempirical equation

$$2 \log \gamma_{\pm} = -A \sqrt{(I)} + 2bI \tag{7}$$

where I, in this case, is given by

$$I = m_{H^+} + m$$

For aqueous solution the constant A is unity and substituting from equation (7) into equation (6) we have

$$\log k - \sqrt{(I)} = \log K - 2bI \tag{8}$$

A plot of $(\log k - \sqrt{I})$ vs. I should thus give a straight line of intercept $\log K$.

Unfortunately, the expression for k, equation (5), contains the true hydrogen ion concentration m_{H^+}, and it is only possible to compute the apparent hydrogen ion concentration m'_{H^+} from the experimental observations. It is consequently necessary to state a further relationship for K so that

$$K = \frac{\gamma'_{H^+}\,\gamma'_{A^-}\,m'_{H^+}\,m'_{A^-}}{\gamma_{HA}\,m_{HA}} \tag{9}$$

or

$$K = \gamma'^2_{\pm} \cdot \frac{m'^2_{H^+}}{M - m'_{H^+}} = \gamma'^2_{\pm}k' \tag{10}$$

The primed quantities in equations (9) and (10) are the apparent values of the quantities concerned corresponding to m'_{H^+}.

In contrast to $\gamma_{\pm}$, $\gamma'_{\pm}$ is not equal to unity at infinite dilution and k' does not approach K as the ionic strength decreases. Denoting the limiting values of $\gamma'_{\pm}$ and k' as $\gamma^0_{\pm}$ and k^0 at infinite dilution

$$K = (\gamma^0_{\pm})^2 k^0 \tag{11}$$

Moreover, it can be shown that when using apparent values, the relationship which corresponds to equation (8) may be stated as

$$\log k' - \sqrt{(I')} = \log k^0 - 2bI' \tag{12}$$

The quantities k' and I' may be calculated from the experimental observations, and hence equation (12) may be used to determine $\log k^0$. The ionic strength term in equation (12) is an apparent value, as it includes a contribution from the apparent hydrogen ion concentration.

If the e.m.f. of cell A is measured for various concentrations of sodium chloride and a constant concentration of acid, k' may be calculated for the various values of I'. A plot of $(\log k' - \sqrt{I'})$ vs. I' will give a straight line of intercept $\log k^0$.

If the above series of measurements is repeated at different concentrations of acid, several values of $\log k^0$ will be obtained, each one corresponding to a particular concentration of acid. These values of k^0 may now be used to determine K since k^0 and K are related by the equation

$$\log \frac{k^0}{K} = 2SM$$

where S is a semiempirical constant.

A plot of log k^0 vs. M thus gives a straight line of intercept log K.

Apparatus and Chemicals

Cambridge slide wire potentiometer with accessories, thermostat, special glass cell, platinum electrode, silver–silver chloride electrode, quinhydrone, 0·01 M hydrochloric acid, chloroacetic acid, sodium chloride.

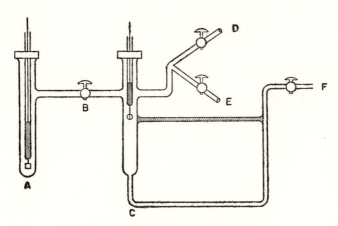

FIG. 1. Glass cell with electrodes.

Method

The cell is rinsed successively with water, distilled water, redistilled alcohol and is then dried. It is placed in a thermostat and a small quantity of quinhydrone is dropped into compartment A (Fig. 1). The platinum electrode is inserted into compartment A and the silver–silver chloride electrode in compartment C.

The compartment C is evacuated by means of a water pump attached at F. Air is subsequently allowed to enter the inlet E. The compartment C is re-evacuated and filled with 0·01 M hydrochloric acid solution to the level of the silver–silver chloride electrode. The acid is sucked out of compartment C. The stop-cock B is opened, and the entire cell evacuated.

The cell is then filled with 0·01 N hydrochloric acid and the electromotive force (E_B) determined. This solution is removed with the aid of the water pump at F. The cell is subsequently filled with a solution of chloroacetic acid ($M = 0·02$) and sodium chloride ($m = 0·02$), and the electromotive force (E_A) determined. Further measurements are

H*

made with 0·02 M chloroacetic acid, but with increasing concentrations of sodium chloride ($m = 0·05$, 0·1 and 0·2).

The 0·02 M chloroacetic acid is replaced with 0·05 M chloroacetic acid and a new set of values for (E_A) are made with sodium chloride ($m = 0·02$, 0·05, 0·1 and 0·2). Further sets of values are obtained with chloroacetic acid ($M = 0·1$ and 0·2).

The extrapolation function ($\log k' - \sqrt{I'}$) is calculated from $m'_{\mathrm{H}}{}^+$ and each of the groups of values which correspond to different values of M are plotted separately against I', enabling the $\log k^0$ values to be determined (the value of $\log k'$ at $I' = 0$). These values for $\log k^0$ are subsequently plotted against the corresponding values for M, enabling $\log K$ to be determined (the value of $\log k^0$ where $M = 0$).

[1] H. S. HARNED and B. B. OWEN, *J. Amer. Chem. Soc.*, **52**, 5079 (1930); WRIGHT, *Ibid.*, **56**, 314 (1934).

[2] H. S. HARNED and B. B. OWEN, *The Physical Chemistry of Electrolytic Solutions*, 3rd ed., Reinhold (1958).

[3] H. S. HARNED and B. B. OWEN, *Ibid.* (1958).

Dissociation Constant of an Acid
(Spectrophotometric Method)

Discussion

The dissociation constant of a weak acid can be determined spectrophotometrically if there is a wavelength in the spectrum where there is a significant difference in the absorption of the undissociated molecule and that due to the charged anion in solutions of known pH. This method provides an accurate one for the determination of the dissociation constant of a colourless weak acid, e.g. benzoic acid which absorbs light in the ultraviolet region. If A_{HA} is the absorbance in an extremely acid solution, where the absorption may be considered to be due solely to the undissociated molecule, and A_A is the absorbance of an extremely alkaline solution where the absorption is due solely to the anion, then the absorbance A, of a solution of intermediate pH is given by

$$A = (1-\alpha)A_{HA} + \alpha A_A$$

Hence

$$\alpha = \frac{(A_{HA}-A)}{(A_{HA}-A_A)} \tag{1}$$

where α is the degree of dissociation.

The thermodynamic dissociation constant K_a is given by the equation

$$K_a = \frac{a_{H^+} \cdot a_{A^-}}{a_{HA}}$$

$$= \frac{m_{H^+} \cdot m_{A^-}}{m_{HA}} \cdot \frac{\gamma_{H^+} \cdot \gamma_{A^-}}{\gamma_{HA}}$$

where m is the molality and γ is the activity coefficient. Hence

$$pK_a = pH - \log\frac{m_{A^-} \cdot \gamma_{A^-}}{m_{HA} \cdot \gamma_{HA}}$$

If the ionic strength of the medium is not too high, the activity coefficient of the undissociated molecule may be taken as unity, hence

$$pK_a = pH - \log\frac{\alpha}{1-\alpha} - \log\gamma_{\pm} \qquad (2)$$

assuming that $\gamma_{A^-} = \gamma_{\pm}$, the mean ionic activity coefficient of the acid.

Hence from equations (2) and (1)

$$pK_a = pH - \log\frac{A_{HA} - A}{A - A_A} - \log\gamma_{\pm} \qquad (3)$$

or

$$pK_a = pH - \log\frac{\epsilon_{HA} - \epsilon}{\epsilon - \epsilon_A} - \log\gamma_{\pm} \qquad (4)$$

where ϵ is the molar absorptivity. Robinson has suggested[1] an empirical equation (a modification of the Debye–Hückel relationship) to enable $\gamma_{\pm}$ at 25°C to be calculated

$$-\log\gamma_{\pm} = 0.5092\sqrt{(I)}/(1+\sqrt{I}) - 0.2\ I \qquad (5)$$

where I is the total ionic strength $\frac{1}{2}\Sigma c_i z_i^2$, where z_i is the valency of the ion and c_i is expressed as mole/l.).

Apparatus and Chemicals

Spectrophotometer, e.g. Unicam SP.800, Hilger Uvispek, constant temperature cell holder, e.g. Adkins Thermostatted Cell Holder and Controller, 5×50 ml graduated flasks, pH meter with electrode system, e.g. Pye potentiometric pH meter (11088), thermostat, 0·001 M solution of benzoic acid or stock solution of other acid, 0·1 N hydrochloric acid, 0·1 N sodium hydroxide solution.

Method

The absorbances are best measured at a wavelength where the molar absorptivities of the undissociated acid and the anion differ most. If this is not known it may be determined in the following way. Solutions of 10^{-4} M benzoic acid in 0·01 N sodium hydroxide and 10^{-4} M benzoic acid in 0·01 N hydrochloric acid are prepared from the stock solutions. The absorbances of these solutions are determined over the whole spectrum. (A recording spectrophotometer, e.g. the Unicam SP.800, is convenient for this purpose.)

If the pH values of these solutions are not sufficiently extreme to displace the equilibrium

$$HA \rightleftharpoons H^+ + A^-$$

completely to the left in the acid solution and completely to the right in the alkaline solution, the observed absorbances may be due to a mixture of HA and A⁻. In order to ascertain that the observed absorbances are due solely to the undissociated molecule in the acid solution and the anion in the alkaline solution, the measurements are repeated with 10^{-4} M solutions of benzoic acid in both 0·1 N sodium hydroxide and 0·1 N hydrochloric acid. If there is no change greater than 1 per cent in the intensity of any peak in the spectra it may be assumed that the acid solution spectrum is due solely to HA and the alkaline solution spectrum is due solely to A⁻. If there is a change greater than 1 per cent the measurements must be repeated on increasingly acidic and alkaline solutions until consistent absorbances are obtained. Examination of the two spectra will reveal the most suitable wavelength for the experimental observations.

The absorbances for HA and A⁻ should be measured accurately using a manual spectrophotometer, e.g. a Hilger Uvispek, at the wavelength indicated by the above observations made with the recording instrument. The absorbance of a buffer solution of benzoic acid in a buffer solution of intermediate pH is measured at the same wavelength. For the most accurate results the pH of the buffer solution should be as near as possible to the pK_a of the acid.[2,3]

If time is limited it may be assumed that for benzoic acid solutions, absorbances should be measured at 240 mμ,[4] and that the pH of the buffer solution should be 4·0. A slight excess of acetic acid in a mixture of 0·01 M acetic acid and 0·01 M sodium acetate gives a pH of approximately 4·0.

Five millilitre portions of the stock solution of benzoic acid are added to each of three 50 ml flasks. One solution is made up to 50 ml with 0·1 N hydrochloric acid, another is made up with 0·1 N sodium hydroxide and the third is made up with the buffer solution. The pH of this last solution at the temperature of the experiment is measured with a pH meter. The absorbance of each solution is measured at 240 mμ.

The pK_a of the acid is calculated using equations (3) or (4), and converted to a value of K_a.

The experiment is repeated several times and an average value of K_a is calculated.

[1] R. A. ROBINSON, *The Structure of Electrolytic Solutions*, edited by W. J. Hamer, J. Wiley, 1959.

[2] A. ALBERT and E. P. SERJEANT, *Ionization Constants of Acids and Bases*, Methuen, 1962.

[3] G. KORTÜM, W. VOGEL and K. ANDRUSSOW, *Dissociation Constants of Organic Acids in Aqueous Solution*, Butterworths, 1961; J. M. WILSON, N. E. GORE, J. E. SAWBRIDGE, F. CARDENAS-CRUZ, *J. Chem. Soc.*, **B**, 852 (1967).

[4] R. A. ROBINSON and A. I. BIGGS, *J. Aust. C.S.*, 1956, **10**, 128 (1957).

Dissociation Constant of a Weak Acid
(Accurate Conductimetric Method)

Discussion

For the exact determination of a dissociation constant, activities rather than concentrations, and the consequences of the Debye–Hückel theory rather than the Arrhenius theory must be considered. The thermodynamic dissociation constant of an acid is given by

$$K_a = \frac{a_{H^+} \cdot a_{A^-}}{a_{HA}} = \frac{c_{H^+} \cdot c_{A^-}}{c_{HA}} \cdot \frac{f_{H^+} \cdot f_{A^-}}{f_{HA}} \tag{1}$$

where f_{H^+} and f_{A^-} are the activity coefficients of the ions in terms of molarities and f_{HA} is the activity coefficient of the undissociated acid. If the ionic strength of the medium is not too high, f_{HA} never differs greatly from unity and we may write

$$K_a = \frac{\alpha^2 c \cdot f_\pm^2}{(1-\alpha)} \tag{2}$$

where $f_\pm$ is the mean ionic activity coefficient.

The degree of dissociation can be determined from conductance data for the weak acid[1] using a modified form of the Onsager equation. The degree of dissociation is given by

$$\alpha = \frac{\Lambda}{\Lambda_0} \cdot F \tag{3}$$

where Λ is the equivalent conductance of the solution, Λ_0 the equivalent conductance at zero concentration and F is given by

$$F = 1 + \left\{ \frac{B_1 \Lambda_0 + B_2}{\Lambda_0^{3/2}} \right\} \sqrt{(\Lambda c)} \tag{4}$$

where $B_1 = 0\cdot230$, $B_2 = 60\cdot65$ and c is the concentration in equivalents per litre. Combining equations (3) and (2) we have

$$\Lambda F = \Lambda_0 - \frac{F^2 f_\pm^2 \Lambda^2 c}{K_a \Lambda_0} \tag{5}$$

224

The activity coefficient $f_{\pm}$ can be calculated from

$$\log f_{\pm} = -A\sqrt{(c\alpha)} = -A\sqrt{(c\Lambda F/\Lambda_0)} \tag{6}$$

where $A = 0 \cdot 509$ in water at $25°C$ for uni-univalent electrolytes. Equation (5) may be rewritten

$$\Lambda F = \Lambda_0 - \frac{V}{K_a\Lambda_0}$$

where

$$Z = F^2 f_{\pm}^2 \Lambda^2 c$$

Hence, using a provisional value of Λ_0 to calculate F as defined in equation (4), a plot of ΛF against Z can be prepared. This graph should be linear and when extrapolated to zero give a more accurate value for Λ_0. The function F may now be recalculated from equation (4) using this improved value of Λ_0. This process is repeated until consistent results for Λ_0 are obtained. The thermodynamic dissociation constant may then be calculated from the slope of the final graph which is equal to $-1/(K_a\Lambda_0)$.

Apparatus and Chemicals

Conductivity bridge assembly, conductivity cell suitable for a weak electrolyte, 25 ml pipette, 50 ml graduated flask, conductivity water, $0 \cdot 1$ D potassium chloride, $0 \cdot 4 \times 10^{-3}$ M 2:4:6 trimethyl benzoic acid.

Method

The conductivity cell is rinsed three times with the solution of potassium chloride. The cell constant is determined.

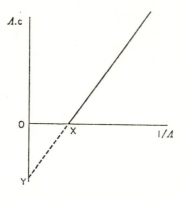

Fig. 1. Plot of $\Lambda \cdot c$ vs. $1/\Lambda$.

The cell is emptied, rinsed with conductivity water and subsequently rinsed three times with the acid solution. A measured amount of the acid solution is added to the conductivity cell. Sufficient solution should be added to enable the electrodes to be immersed at least 1 cm below the surface of the solution. The conductance of the solution is determined.

Solutions of 2:4:6 trimethyl benzoic acid (0.35×10^{-3} M, 0.3×10^{-3} M, 0.25×10^{-3} M, 0.2×10^{-3} M) are prepared by dilution with conductivity water and the conductance determined at each dilution.

A plot of Λc against $1/\Lambda$ is prepared (see Fig. 1).

The value of F is calculated as defined in equation (4) using a provisional value of Λ_0, say $1/(OX)$ as shown in Fig. 1. A plot of ΛF against Z is prepared. This plot is extrapolated to zero to give a more accurate value of Λ_0. The function F is recalculated using this improved value of Λ_0. A second plot of ΛF against Z is prepared and K_a calculated from the slope.

[1] T. SHEDLOVSKY, *Physical Methods of Organic Chemistry*, edited by A. Weissberger vol. 1, chap. xxv, Interscience Publishers, New York.

Dead-stop End Point Titration Technique

Discussion

In an amperometric titration the mercury drop in the polarographic cell is held at a definite potential while the diffusion currents throughout a given titration are recorded. At the end point there is an abrupt change in the diffusion current value.

A simplified experimental modification of this method is the "dead-stop" end point technique, where the two electrodes, usually smooth platinum, are immersed in the solution to be titrated, and a small potential difference, 15–400 mV, depending on the required sensitivity, is set up between these two electrodes. This method is of particular simplicity in the case of the iodine thiosulphate titration since at the end point the current decreases to an almost negligible value (see Fig. 1). This method was at first thought to be only applicable to the iodine thiosulphate system. Recent work has, however, shown the technique not to be so dependent on the diffusion current theory as in amperometric titrations, but rather on whether the redox system between the electrodes is a reversible or an irreversible one. The reversible system permits current flow through the cell, and the irreversible system virtually stops the conduction process. Other possible systems for study by this method are the titration of: (1) ferrous salts with ceric salts, (2) metavanadate with ferrous, (3) ferrocyanide with ceric.

This iodine/thiosulphate titration is probably the most precise and sensitive method known for iodine estimations. As long as free iodine remains in the solution the main electrode process can be regarded as the iodine/iodide reaction acting in its reversible capacity at each platinum electrode. Immediately the end point is reached, however, the last of the free iodine is removed and the iodine/iodide reaction can no longer occur. Also since the thiosulphate/tetrathionate reaction is highly irreversible, no detectable current will be observed.

Apparatus and Chemicals

Circuit assembly as illustrated in Fig. 2 including a sensitive microammeter, and millivoltmeter, nominal 0·1 M iodine solution and 0·1 M sodium thiosulphate solution, platinum wire electrodes, magnetic stirrer.

Method

The platinum electrodes are immersed in the cell containing 25 ml of the iodine solution, and it should be ensured that the electrodes are fully immersed by the addition of sufficient distilled water. The rheostat R is adjusted so that a current of about 200 μA passes. A preliminary titration is then performed to verify that the cell conditions and applied voltage are of the correct order for the production of a satisfactory "dead-stop" end point. The solution should be stirred continuously throughout the titration. If necessary the current is adjusted and a few accurate titrations are performed.

A graph of current readings against volume of added sodium thiosulphate solution should be plotted (see Fig. 1). The electrometric end point should be compared with the normal volumetric colour change end point, and the molarity of the iodine solution calculated.

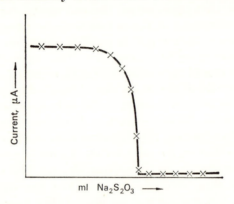

FIG. 1. Variation of current with added volume of sodium thiosulphate.

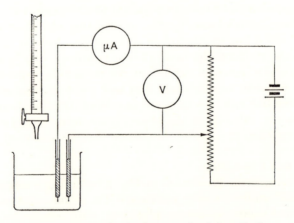

FIG. 2. Circuit for dead-stop end point method.

Potentiometric Titration—Solubility of Silver Halides

Discussion

When a solution of a soluble metallic halide, other than fluoride, is titrated with silver nitrate solution, silver halide is precipitated. If a silver electrode is immersed in the solution during the titration its potential is given by

$$E_{Ag} = E^0_{Ag} + \frac{RT}{F} \ln a_{Ag^+} \tag{1}$$

where a_{Ag^+} is the activity of the silver ions in solution.

At the equivalence point

$$a_{Ag^+} = a_{x^-}$$

where a_{x^-} is the activity of the halide ions. The solubility product K_{sp} is given by the relationship

$$a_{Ag^+} \times a_{x^-} = K_{sp}$$

hence $a_{Ag^+} = \sqrt{K_{sp}}$ at the equivalence point. Substituting in equation (1)

$$E_{Ag} = E^0_{Ag} + \frac{RT}{F} \ln \sqrt{K_{sp}}$$

or

$$E_{Ag} = E^0_{Ag} + \frac{RT}{2F} \ln K_{sp} \tag{2}$$

If the silver electrode is combined with a saturated calomel electrode, E_{Ag} can be determined from the e.m.f. of the cell as

$$E_{cell} = E_{Ag} - E_{cal} \tag{3}$$

Hence, knowing the standard electrode potential of silver, E^0_{Ag} and the potential of the saturated calomel electrode E_{cal}, the solubility product of the silver halide can be calculated by the application of equations (2) and (3).

To carry out a potentiometric titration the actual electrode potential is not required, it is sufficient to record the e.m.f. of the complete cell

E_{cell}, after the addition of known amounts of silver nitrate solution. The end point can be determined graphically by plotting E_{cell} vs. v ml of silver nitrate added.

If a mixture of metallic halides is titrated with silver nitrate solution, the more insoluble silver halide will be precipitated first. As soon as some of the silver halide is precipitated the solution will be saturated with respect to this particular halide, and its solubility will determine the activity of the silver ions in solution. This activity will increase slightly as precipitation continues owing to the consequent decrease in the chloride ions. As the precipitation of the silver electrode is governed by the activity of the silver ions in solution the potential of the silver electrode will also increase slowly during the precipitation.

When the more insoluble of the two silver halides has been completely precipitated, the next drop of silver nitrate to be added will start to precipitate the second of the silver halides. The solution will now be saturated with the second silver halide and the activity of the silver ions will increase sharply to a new value which will suffer a further slight increase during the precipitation of the more soluble halide. A step will thus occur in the E_{cell} vs. ml graph.

When the second silver halide has been completely precipitated, a further increase in the activity of the silver ions will occur due to the presence of excess silver nitrate. This will be reflected in a corresponding increase in the potential of the silver electrode.

Apparatus and Chemicals

Potentiometer with accessories, calomel electrode, silver electrode, automatic stirrer, 0·1 N silver nitrate, solution of sodium chloride and sodium iodide (*ca.* 0·1 N with respect to each).

Method

A known volume (10 ml) of the mixture of the metallic halides is placed in a beaker. The electrodes are attached to the potentiometer, the silver electrode is immersed in the mixed halide solution and connection is made to the calomel electrode through an ammonium nitrate salt bridge. The solution is diluted with distilled water so that the level covers the electrodes. The stirrer is placed in solution and set in motion. The mixture of the metallic halides is titrated with the silver nitrate, the e.m.f. (E_{cell}) of the cell being measured and recorded after the addition of each portion of the silver nitrate.

A plot of E_{cell} vs. v ml of silver nitrate is drawn. The values of E corresponding to: (a) the precipitation of silver iodide, (b) the precipitation of silver chloride are noted. The solubility products of the silver halides are calculated by the application of equations (3) and (2).

The Polarographic Method of Analysis

Discussion

Nernst suggested that a sensitive form of chemical analysis was possible using a small polarized electrode connected in series with a large non-polarized electrode. Heyrovsky[1] used a dropping mercury electrode and a mercury pool, similar to the apparatus shown in Fig. 1, for the analysis of dilute zinc sulphate and cadmium sulphate solutions. By continuously polarizing the mercury droplets, he obtained a plot of current vs. applied voltage as shown in Fig. 2. The curve heights and half wave potentials (H.W.P.) are characteristic of the quantitative and qualitative aspects respectively of the species under test. This curve height is controlled by the diffusion current, the original Ilkovic[2] equation for which is

$$i = A z D^{1/2} m^{2/3} c t^{1/6}$$

where i is the instantaneous diffusion current, A a constant, D the diffusion coefficient of the species under test, m the mass of mercury flowing per second, t the instantaneous time, c the concentration of the species under test, and z the number of electrons involved in the reduction of one molecule. The half wave potential is given as a first approximation by

$$E_{1/2} = E^0 - \frac{RT}{zF} \ln \frac{D^{1/2}_{\text{oxidant}}}{D^{1/2}_{\text{reductant}}}$$

where D_{oxidant} and $D_{\text{reductant}}$ are the diffusion coefficients of the species in the oxidized and reduced forms. This equation indicates that the half wave potential $E_{1/2}$ should only differ from the standard reversible redox potential, E^0, by a negligibly small value, if the diffusion coefficients of the oxidant and reductant are almost equal.

As shown in Fig. 2 for the initial application of a small voltage there is a small current which remains almost unchanged (AB) with increasing voltage until the decomposition potential of one of the ionic species under test is reached when there is a relatively large current rise. The constant half wave potential term was introduced by Heyrovsky, in place of the variable decomposition potential value. This sudden surge in current (BC) due to the discharge of the ions of the test species at the cathode is followed by a limiting value of the current (*CD*). Depletion of the reacting ions at the cathode interface has now occurred, and this limiting current is controlled by the rate of diffusion of the ions from the bulk solution. Further increase in applied e.m.f. will cause no marked change in the value of this limiting current until the decomposition potential of a new ionic species is reached. In the polarogram shown in Fig. 2 the cadmium ion step is followed by that

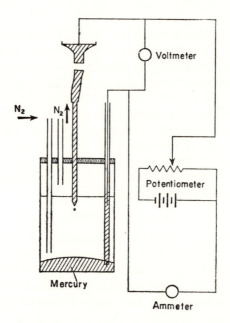

Fig. 1. Apparatus for polarographic analysis.

of the zinc ion. This analysis can be carried out in a deoxygenated base solution of 0·1 M potassium chloride which may require a small amount of gelatine to maintain the curve shape as shown.

The object of this investigation is:

(a) to show that the current rise produced by a particular ion is proportional to the concentration of that ion;

(b) to determine the H.W.P. values for cadmium and zinc.

Apparatus and Chemicals

Cambridge manual polarograph, galvanometer, accumulators, 0·1 M potassium chloride, gelatine, sodium sulphite, cadmium sulphate (A.R.) and zinc sulphate (A.R.).

Method

Before connecting the polarograph to the accumulator and the galvanometer, the instructions supplied by the manufacturer should be carefully read.

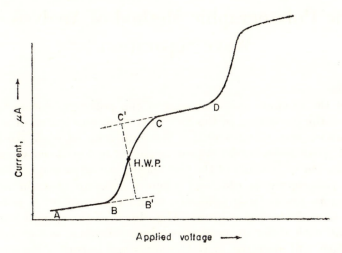

FIG. 2. Cadmium and zinc polarograms.

A beaker is filled with 100 ml of distilled water, to which 0·4 g of cadmium sulphate and 0·4 g of zinc sulphate is added. Five millilitres of this solution is then transferred to a graduated flask and made up to 1 l. with 0·1 M potassium chloride solution. Approximately 5 ml of this final solution is then poured into the polarographic cell and any dissolved oxygen removed by bubbling nitrogen through the solution or by adding a small amount of the powdered sodium sulphite. The drop rate of the mercury cathode should be one drop every 3–5 sec. The polarograph is switched on and the current recorded as the applied voltage is increased in steps of 0·1 V.

Twenty-five millilitres of the original solution of potassium chloride, zinc sulphate and cadmium sulphate is then diluted with an equal volume of distilled water. Approximately 5 ml of this diluted solution is then added to the polarographic cell, deaerated and the polarogram plotted. Both polarograms are interpreted as shown in Fig. 2, the relevant H.W.P. values are determined and the proportionality between the wave heights and concentrations of the corresponding ions is checked.

[1] J. HEYROVSKY, *Chem. Listy*, **16**, 256 (1922).
[2] D. ILKOVIC, *Coll. Czech. Chem. Communs.*, **6**, 498 (1934).

The Polarographic Method of Analysis— Wave Separation

Discussion

One of the obvious disadvantages of the polarographic method of analysis is the possibility of the half wave potential values of any two species under test being within a few millivolts of each other. The resultant polarogram would appear as shown in Fig. 1, curve (a), and a simultaneous estimation of the two species would be impossible. This type of polarogram is obtained when a mixture of cadmium and thallium ions are polarographically examined in an 0·1 M potassium chloride base electrolyte solution. The object of this experiment is to show that with different base electrolyte solutions, complexing will occur which will alter the H.W.P. values and permit a simultaneous analysis of both cadmium and thallium.

Apparatus and Chemicals

Cambridge pen recording polarograph, 0·1 M potassium chloride (a), solution of equal parts of 0·5 M ammonium hydroxide and 0·5 M ammonium chloride (b), 0·1 M potassium hydroxide (c), thallous sulphate, and cadmium sulphate.

Method

Three solutions are prepared for analysis by dissolving 0·032 g of thallous sulphate and 0·021 g of cadmium sulphate in three separate 100 ml portions of the base solutions a, b and c. In this way three solutions of 10^{-4} M thallous sulphate and 10^{-4} M cadmium sulphate are prepared.

The polarograph is then set up, making sure that the drop rate is between 3 and 5 sec. The three solutions are then electrolysed in turn between $-0·2$ V and -1 V. The resultant polarograms are similar to those illustrated in Fig. 1.

It is noted that the H.W.P. for the thallium is more or less constant for the three base electrolytes, while the cadium H.W.P. value becomes increasingly negative from curve (a) to curve (c). The wave heights of the thallium are also approximately constant in the three base electrolytes. The corresponding wave heights for cadmium decrease from (a) to (c), and in curve (c) the wave is almost completely lost. Hence only curve (b) shows the desired wave separation, and indicates the experimental conditions required for the simultaneous analysis of thallium and cadmium.

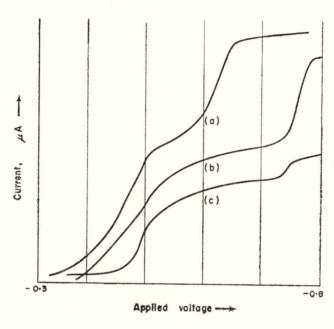

FIG. 1. Polarograms of a cadmium and thallium mixture dissolved in three different base electrolytes.

The nature of the cadmium complexing ability in solutions (b) and (c) should be discussed, together with its possible significance in qualitative analysis.

Polarographic Study of Acetaldehyde

Discussion

Aldehydes give well-defined polarographic steps in a base solution of pH value greater than 7. Acetaldehyde is a particularly good example as it is less hydrated than some of the others and so external factors have less effect on the polarographic step. The highest step is obtained from a 0·1 M lithium hydroxide solution. This step is quantitative for solutions that are less than 0·001 M with respect to acetaldehyde, and for low acetaldehyde concentration the best sensitivity is given by a solution whose pH value is between 12 and 13.

Apparatus and Chemicals

Polarograph, acetaldehyde and 0·1 M lithium hydroxide solution.

Method

An approximately 0·2 M solution of acetaldehyde in 0·1 M lithium hydroxide base solution is made by weighing accurately about 1 g of acetaldehyde and dissolving it in 100 ml of base solution. Solutions containing approximately 2×10^{-4}, 4×10^{-4}, 6×10^{-4}, 8×10^{-4} and 10×10^{-4} M acetaldehyde in the base solution are made up by suitably diluting aliquots of the original solution with the base solution. Samples of these diluted solutions are transferred to the polarographic cell and the polarographic step is recorded without removing dissolved oxygen. This procedure is then repeated with a portion of the base solution only. The half wave potential of acetaldehyde in 0·1 M lithium hydroxide is determined from the polarograms and the step heights are plotted against the concentrations of the solutions. The concentration of a solution of unknown molarity of acetaldehyde is obtained from the calibration graph.

Steps can also be produced in the vicinity of the acetaldehyde step by other members of this homologous series—propionaldehyde, n- and iso-butyraldehyde, n-valeraldehyde, n-capronaldehyde and n-heptaldehyde, and for a well-defined acetaldehyde step, therefore, these must not be present.

Preface to Experiments 88 and 89

Safety Precautions to Be Observed when Carrying Out Radiochemical Experiments

The work should be carried out in a radiochemistry laboratory.

Any health hazard which may arise from radioactivity can be reduced to a minimum if common sense precautions are taken.[1] A number of books[2,3] have been written on radiochemical techniques and the student is advised to familiarize himself with the fundamentals before using radioactive tracers.

To reduce the general risk the amount of radioactive tracer should be kept to a minimum consistent with accurate measurement.

External Exposure

External exposure to a radioactive source can be reduced by distance (inverse square law) and by shielding. When working with trace levels ($\sim 1\,\mu c$) the danger from external exposure is negligible; nevertheless, sources should always be handled with tweezers and tongs and the time spent in their actual handling should be kept as short as possible.

Ingestion and Inhalation

Ingestion and inhalation may present a real hazard when working with open sources, however small. It is therefore essential not to eat, drink or smoke in the laboratory. No mouth operated pipettes should be used. Cosmetics should not be applied in the laboratory. Whenever possible work with radioactive materials should be carried out in a fume cupboard. If there is a choice, operations in solution are to be preferred to those in which the radioactive material is a dry powder.

Contamination

Contamination and spread of radioactivity must be controlled. It is advisable to divide the laboratory into active and inactive areas. Care must be taken not to allow any radioactive material to come into contact with skin or clothing. Contamination of the hands can be avoided by the use of rubber gloves so that the skin is never in contact with the active side. Gloves and hands should be checked after each operation by means of a radiation monitor. Experiments should not be written up in an "active" area. If contamination does occur it should be cleaned up at once (by washing and rinsing into the sink) and the area, hands, etc., should be subsequently monitored for any residual

237

activity. By working in trays, spread of contamination can be limited to the immediate area of a spill. It should be realized that this extreme cleanliness of working with radioactive materials is not only necessary because of the health hazard, but also because unwanted traces of activity would lead to inaccurate results.

Waste Disposal

Waste disposal must be carried out with external and internal exposure in mind. Records should be kept of all radioactive material dispensed and disposed of. Accumulation of radioactive waste in the laboratory is undesirable. Solutions should be flushed down the sink so that the isotope is present below its maximum permissible level in drinking water. Solid waste can be collected in small plastic bags, which are sealed and disposed of with the general solid waste.

[1] J. W. LUCAS, Radiation safety and health physics, *Nature, Lond.*, **184**, 329 (1959)
[2] R. T. OVEVMAN and H. M. CLARK, *Radioisotope Techniques*.
[3] R. A. FAIRES and B. H. PARKES, *Radioisotope Laboratory Technique*.

Separation of Radioelements by Ion Exchange

Discussion

In using radioactive tracers to investigate chemical reactions care must be taken that the isotope employed is radiochemically pure. The presence of more than one radionuclide could lead to the misinterpretation of results. This situation can occur when a radioactive isotope has a short-lived "daughter" which is responsible for (a) the presence of a particular type of radiation or (b) a fraction of the total activity. For example, caesium-137 is known as a gamma emitter with a half life of 30 years. To use this isotope as a tracer for caesium or other univalent cations would be incorrect, because, in reality, it is a short-lived "daughter" barium-137m which is responsible for the 0·66 meV gamma radiation. The decay scheme is:

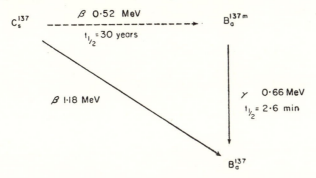

The presence of the short-lived "daughter" can be shown by any process which separates monovalent and divalent metal ions. This separation can therefore be achieved by passing a caesium-137 solution over a cation exchange resin and "milking" the barium off with a complexing agent.

After the separation the short half-life of the "daughter" isotope barium-137m can be calculated.

Apparatus and Chemicals

Ambelite resin IR-120, 2 M hydrochloric acid, 2 M sodium hydroxide, 0·1 M ethylene diaminetetra-acetic acid (EDTA) in 1 M sodium hydroxide, stop watch, caesium/barium carrier solution ($\sim 10^{-3}$ M),

239

caesium-137 tracer solution (1 drop ~ 0·05 ml ~ 0·1 μc), rubber gloves, tray, separate active and inactive glassware, safety pipettes and liquid Geiger–Müller counter (a description of this counter, ancillary electronic equipment and counting technique can be found in Refs. 2 and 3 of the Safety Precautions given immediately preceding this experiment).

Method

Before commencing this experiment read carefully the safety instructions which are given at the beginning of this section.

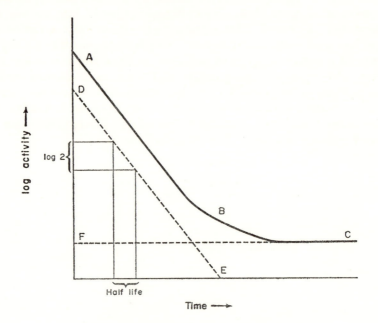

FIG. 1. Variation of log activity with time.

Approximately 5 ml of IR-120 resin is placed in a burette and washed in sequence with: (i) 7·5 ml of 2 M hydrochloric acid, (ii) distilled water until neutral, (iii) 10 ml of 2 M sodium hydroxide, and (iv) distilled water until neutral. The column is most efficiently washed and operated if solutions are allowed to run so that the level of liquid coincides with the top of the resin before the next addition of fluid.

A drop of radioactive caesium-137 (0·10 μc) is added to 5 ml of caesium chloride (barium contaminated) carrier solution. The labelled solution is poured over the column and washed through with 5 ml of distilled water. The barium-137 m is now "milked" off the column with 5 ml of 0·1 M ethylene diaminetetra-acetic acid (EDTA) in 1 M sodium hydroxide and two 5 ml portions of distilled water. As the complexed barium starts to come off the column the stop watch is started. Ten millilitres of the eluate are transferred to the liquid Geiger–Müller tube and counted for 30 sec at 1 min intervals. A graph is plotted of log activity vs. time (Fig. 1).

The total activity ABC consists of a rapidly decreasing section DE due to the Ba^{137m} and a fairly constant section FC due to the long-lived isotope Cs^{137} which exchanges with the sodium ions in the EDTA solution.

If the level of activity represented by FC is subtracted from the curve ABC the resultant curve DE corresponds to the activity of the Ba^{137m}.

The half life of the "daughter" isotope is calculated by taking any two points on the curve DE, the ordinate values of which differ by log 2; the half life is then equal to the differences in the values of the abscissae of the two points.

Distribution Constant by a Radioactive Tracer Technique

Discussion

Dryssen (1955)[1,2] has deduced that when strontium is extracted as its oxinate from an aqueous solution into chloroform it goes in the form of a complex, viz.

i.e.

$$SrOx_2[HOx]_2$$

This is deduced from the fact that a single curve results from plotting

$$\left[\log \frac{[Sr]_{\text{total org.}}}{[Sr]_{\text{total aq.}}} - 2\log[HOx]_{\text{org.}}\right]$$

against $\log[Ox^-]$.

Equilibrium between the ion and complex can therefore be written as

$$Sr^{2+} + 4HOx \overset{pH10}{\rightleftharpoons} SrOx_2 . 2HOx + 2H^+$$

so that the net distribution ratio (k) is given by

$$k = \frac{[\text{Sr}]_{\text{total org.}}}{[\text{Sr}]_{\text{total aq.}}} = \frac{[\text{SrOx}_2 2\text{HOx}]_{\text{org.}}}{[\text{Sr}^{2+}] + [\text{SrOx}_2] + [\text{SrOx}_2 2\text{HOx}]_{\text{aq.}} + \text{other possible species}}$$

Determination of the total strontium content in each phase can easily be carried out by using a radioactive isotope of strontium. The ratios of the radioactivity in the two phases will be the same as the ratios of the total strontium concentrations, i.e.

$$k = \frac{[\text{Sr}]_{\text{total org.}}}{[\text{Sr}]_{\text{total aq.}}} = \frac{\text{radioactivity in organic layer}}{\text{radioactivity in aqueous layer}}$$

The advantage of the radioactive method is that the complexing agent does not interfere with strontium determination, nor do the different types of species in the aqueous phase in which the strontium might be bound.

The method therefore reduces simply to adding some strontium-89 (half life = 54 days; energy $\beta = 1 \cdot 48$ meV) as a tracer to a strontium solution, adjusting to pH 10 with sodium hydroxide solution and then extracting with oxine in chloroform solution until equilibrium has been established. The solvent and aqueous layers are counted in a liquid Geiger–Müller counter and the net distribution coefficient determined.

Apparatus and Chemicals

Extraction apparatus (Fig. 1), $7 \cdot 57 \times 10^{-5}$ M Sr^{2+} solution, 2 M sodium hydroxide, 1 M 8-hydroxyquinoline in chloroform solution, universal pH paper, Sr^{89} tracer solution (1 drop $\sim$ 0·05 ml $\sim$ 0·1 μc), rubber gloves, tray, separate active and inactive glassware, safety pipettes and liquid Geiger–Müller counter (a description of this counter, ancillary electronic equipment and counting technique can be found in Refs. 2 and 3, p. 238).

Method

Before commencing this experiment read carefully the safety instructions which are given on p. 237.

To 15 ml of a $7 \cdot 57 \times 10^{-5}$ M Sr^{2+} solution 2 M sodium hydroxide is added drop by drop until the solution is pH 10 as shown on a universal indicator paper.

To the alkaline solution (now transferred into the active area) one drop of strontium-89 solution is added and thoroughly mixed. Ten millilitres of this solution are transferred to a liquid Geiger–Müller tube and the radioactivity measured in counts per minute (cpm). A

I

total of approximately 10,000 counts is made to achieve statistical accuracy (see Refs. 2 and 3, p. 238).

Fifteen millilitres of labelled strontium solution is transferred to the extraction apparatus and 15 ml of 1 M oxinate in chloroform solution added. The radioactive background count can be measured while the extraction is taking place.

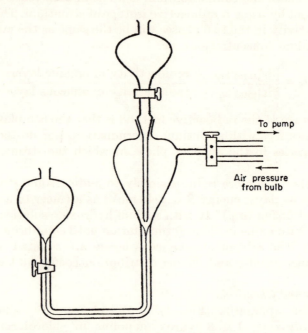

To pump

Air pressure
from bulb

Fig. 1. Extraction apparatus for dealing with radioactive solutions.

The radioactive background count is carried out after carefully cleaning the Geiger–Müller tube with nitric acid, water and finally acetone. True counts per minute will then be given by:

$$\text{cpm}_{\text{corrected}} = \text{cpm}_{\text{active sample}} + \text{correction for} - \text{cpm}_{\text{background}}$$
$$\text{paralysis time}$$

(For paralysis time correction see Refs. 2 and 3, p. 238.)

The extraction of strontium from the aqueous layer in the extraction apparatus (Fig. 1) is carried out by sucking and blowing so that the liquid layers flow through one another. After pulling and pushing the two phases through each other about twelve times, they are allowed to settle and separate. Each layer is then pushed into a separate holding funnel from which 10 ml quantities are transferred by safety pipettes to the Geiger–Müller tube. A count is then taken of the activity in each phase. After correction due to background count and paralysis time the net distribution ratio will be given by

$$k = \frac{\text{cpm}_{\text{corrected}} \text{ in 10 ml of organic phase}}{\text{cpm}_{\text{corrected}} \text{ in 10 ml of aqueous phase}}.$$

[1] D. Dryssen, *Svensk. Kem. Tidskr.*, **67**, 311 (1955).
[2] D. Dryssen, *J. Inorg. Nucl. Chem.*, **8**, 291 (1958).

PART III

PART III

A Chromatographic Study of the Effects of Substitution on Acetophenone

Discussion

The retention volume of a substance injected into a chromatograph will depend upon the degree of interaction of the substance with the stationary phase. The greater the affinity between the two, the greater will be the retention volume. When examining a series of substances such as methyl-substituted acetophenones it is convenient to have a reference substance of a similar nature, such as acetophenone, and express the results as relative retention volumes V_r, given by

$$V_r = V_{N_2}/V_{N_1} \qquad (1)$$

where V_{N_2} is the net retention volume of the substance under investigation and V_{N_1} is the net retention volume of the reference substance in the same column.

The extent of interaction of substituted acetophenones with a stationary phase will depend on the polar qualities of solute and solvent. If both are polar then the greater the polarity the greater will be the degree of interaction. Substituents in the nucleus of acetophenone can affect the polarity of the molecule in different ways. The possible factors to be considered are:

(a) the inductive effect of the substituent group;
(b) the mesomeric effect of the substituent group;
(c) hyperconjugation;
(d) restriction of coplanarity.

As regards this last effect it has already been noted in Experiment 64 that the presence of *ortho*-substituents in acetophenone restricts coplanarity and reduces the contributions of the polar structures II and III.

With sterically hindered ketones of this type there should be less interaction with a polar stationary phase giving rise to lower retention volumes. This decrease in the retention volumes has already been noted for the analogous halogen *ortho*-substituted diphenyls.[1]

Interaction may also arise as a result of the steric properties of the solvent and the solute with respect to each other. This type of interaction will occur whether the solvent is polar or non-polar. It may thus be possible to separate the effects of these factors to some extent by firstly using a polar stationary phase such as dinonyl phthalate and secondly using a non-polar stationary phase such as squalane (2:6:10:15:19:23-hexamethyltetracosane). More information on the polarity of the substituted acetophenones may be obtained from the conjugation in the various compounds as shown by their ultra-violet spectra.

An indication of the interaction between solute and solvent is given by the Raoult law factor as defined by the relationship

$$\gamma = \frac{RTw}{Mp^0 V_N} \qquad (2)$$

where γ is the Raoult law factor and the other symbols have the same significance as defined in Experiment 57.

If the compound under investigation and the reference compound are passed through the same column at the same temperature then by equation (2)

$$\gamma_1 = \frac{RTw}{Mp_1^0 V_{N_1}} \qquad (3)$$

and

$$\gamma_2 = \frac{RTw}{Mp_2^0 V_{N_2}} \qquad (4)$$

where the subscripts 1 and 2 denote the reference compound and the compound under examination respectively. Combining equations (3) and (4) together with equation (1)

$$\frac{\gamma_2}{\gamma_1} = \frac{p_1^0 V_{N_1}}{p_2^0 V_{N_2}} = \frac{p_1^0}{p_2^0 V_r} \qquad (5)$$

From a knowledge of vapour pressures and relative retention volumes the relative Raoult law factor γ_2/γ_1 may be calculated. If these factors for various substituted acetophenones are not very different then the retention volumes must depend largely upon the vapour pressures of the compounds.[1]

Apparatus and Chemicals

Gas chromatograph, spectrophotometer, acetophenone, 2:4-dimethyl acetophenone, 3:4-dimethyl acetophenone, and 2:4:6-trimethyl acetophenone.

The support for the stationary phase in the chromatograph column should preferably be glass beads. Alternatively, the support should be treated with dimethyldichlorosilane vapour to block the polar sites before the stationary phase is applied.

Method

A general discussion of gas–liquid chromatography has already been given in Experiments 56 and 57 and reference should be made to these experiments before proceeding.

Samples of the four ketones are injected into a chromatograph containing a column with dinonyl phthalate as a stationary phase at 100°C. The samples may be injected individually in the first instance to find their approximate retention volumes and then a mixture of the four ketones may be injected to ensure that they are all examined under the same experimental conditions. The results should be expressed as the relative net retention volumes with reference to acetophenone.

The same experiment should be carried out using squalane as the stationary phase and the results obtained with the polar and non-polar solvent are compared.

Given the boiling point data below the vapour pressures of the ketones at 100°C are estimated by a Cox chart method.[2] From the vapour pressure data the ratio p_1/p_2^0 is calculated and hence γ_2/γ_1 for both stationary phases is calculated for each of the substituted ketones. Comment is made upon the results.

BOILING POINT DATA

Acetophenone	202°C at 760 mm
2:4-dimethyl acetophenone	228°C at 760 mm
3:4-dimethyl acetophenone	246°C at 760 mm
2:4:6-trimethyl acetophenone	240·5°C at 735 mm

The ultra-violet spectrum of each substance is examined (Experiment 64).

A report should be written on the effects of substituting methyl groups in the nucleus of acetophenone and these effects should be analysed as far as possible with regard to the polar and steric factors involved.

I*

Which methyl substituted acetophenone should be the next compound to be examined to provide the maximum amount of additional information?

[1] E. A. JOHNSON, The study of steric effects in alkyl phenyls by vapour phase chromatography, *Steric Effects in Conjugated Systems*, p. 174, Proceedings of a Symposium held at the University, Hull, July 1958 by the Chemical Society, Butterworths Scientific Publications (1958).

[2] R. R. DREISBACH, *Pressure–Volume–Temperature Relationships of Organic Compounds*, 3rd ed., Handbook Publishers, Sandusky, Ohio (1952).

A Chromatographic Study of the Relationship between Heats of Solution and Molecular Structure

A. N. STRACHAN,
Loughborough University of Technology

Discussion

The vapour pressure p of a solute above a solution depends on three factors:

(a) the concentration of the solute as measured by its mole fraction N,

(b) the volatility of the solute as measured by its vapour pressure in the pure state p^0

(c) the degree of interaction of the solute with the solvent as measured by the activity coefficient or Raoult's law factor γ and by the heat of solution ΔH. The greater the interaction, the smaller γ and the more negative ΔH. (Note: since heat is invariably evolved when a gaseous solute dissolves, ΔH will have a negative sign.) Thus

$$p = N\gamma p^0 = NA e^{\Delta H/RT} \qquad (1)$$

where A is a constant for a given solute and is another measure of volatility. The more volatile the solute the larger is A.

In gas chromatography, assuming ideal behaviour in the gas phase, the retention volume V_N varies inversely as p and so depends equally on these same three factors It can be shown that:

$$V_N = \frac{RTw}{M\gamma p^0} = \frac{RTw}{MA} e^{-\Delta H/RT} \qquad (2)$$

where w is the weight of stationary phase (solvent) in the column and M is the molecular weight of the stationary phase

Hence
$$\frac{d\log_{10} V_N}{d(1/T)} = \frac{-(\Delta H + RT)}{2 \cdot 303R} \qquad (3)$$

253

RT is small compared with ΔH and, provided the temperature range is not large, a plot of $\log_{10} V_N$ v $1/T$ will be linear. From the slope the heat of solution ΔH can be determined by equation (3). The temperature T on the right-hand side of the equation should be taken as the mean of the temperature range covered.

Since $V_N = jAZ.F$ (equation (1) Experiment 56), if the pressures at the two ends of the column, which determine j, are kept the same at all temperatures, it is then only necessary to measure AZ and F at each temperature and to plot $\log_{10} AZ.F$ against $1/T$.

Apparatus and Chemicals

Gas chromatography, n-butyl alcohol, iso-butyl alcohol, sec-butyl alcohol, tert-butyl alcohol, dinonyl phthalate column, squalane column.

Method

Samples of the four butyl alcohols are injected into the chromotograph containing the dinonyl phthalate column and with the inlet and outlet pressures kept constant, AZ and F are measured for each alcohol at several different temperatures between 80 and 120°C. The heats of solution of the four alcohols in dinonyl phthalate are obtained. The investigation is repeated using the squalane ($C_{30}H_{62}$, 2:6:10:15:19:23-hexamethyltetracosane) column. All eight heats of solution are then compared and an attempt made to explain their differences in terms of the differences in molecular structure of the four alcohols and of the two column liquids. The relative degrees of interaction of other components with various column liquids may be investigated in a similar manner.

Steric Effects and Resonance

Discussion

The presence of alkyl substituents in the *ortho*-position of a series of substituted acetophenones prevents co-planarity between the benzene ring and the carbonyl group. Acetophenone may be represented as a combination of three structures, I, II and III.

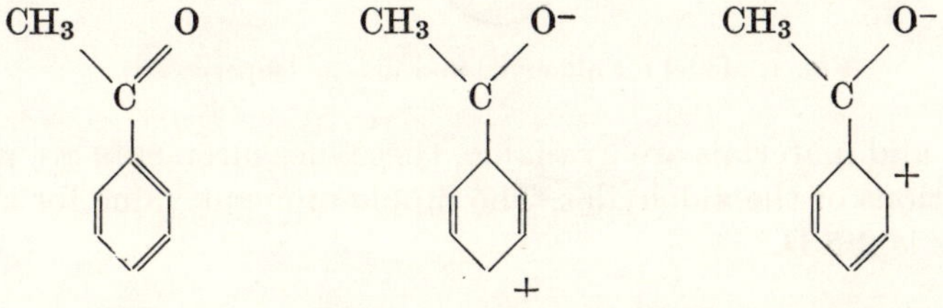

The prevention of co-planarity will reduce the contribution of structures II and III, in which case the dipole moment should be reduced and approach the value for the corresponding aliphatic compound.[1] In *ortho*-substituted benzaldehydes where there is no methyl group attached to the carbonyl group, there will be less steric interference as indicated in Fig. 1. Resonance will not be inhibited and the dipole moment should be higher than that of the aliphatic analogue.

Apparatus and Chemicals

Assembly for the measurement of dipole moments, pyknometer, Abbé refractometer, pure samples of benzene, acetophenone, 2 : 4 : 6-trimethyl acetophenone, benzaldehyde, and 2 : 4 : 6-trimethyl benzaldehyde.[2]

Method

The dielectric constants of a series of dilute solutions of the ketones in benzene are determined. The dielectric constant of the solvent, benzene, is measured.

The density of benzene and of each of the solutions of the ketone in benzene are determined with the pyknometer.

The refractive indices of the ketones are measured.

From the data obtained, the dipole moments of the ketones are calculated by the method given in Experiment 53. The dipole moments obtained are compared with the value for the corresponding aliphatic ketones, 2·7 D.

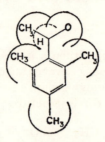

Fig. 1. Model for aldehyde and ketone (superposed).

If time and materials are available, these measurements are repeated with solutions of the aldehydes. The dipole moment value for aliphatic aldehydes is 2·5 D.

[1] R. G. Kadesch and S. W. Weller, *J. Amer. Chem. Soc.*, **63**, 1310 (1941).
[2] E. C. Horning (Ed.), *Organic Synthesis*, vol. 3, p. 549, John Wiley.

Dipole Moments of Polar Compounds (Guggenheim's Method)

A. RICHMOND,

West Ham College of Technology

B. D. COSTLEY,

Wolverhampton College of Technology

Discussion

The method of determining dipole moments described in Experiment 53 involves the accurate determination of the densities of the solutions of the polar solute in the non-polar solvent. Guggenheim[1,2,3] demonstrated how this could be avoided using the relationship

$$\mu^2 = \frac{10^{36} \cdot 9kT}{4\pi N} \cdot \frac{3M_2}{(E_0+2)^2 d_0} \left[\underset{w \to 0}{Lt} \frac{(\epsilon - \epsilon_0)}{w} - \underset{w \to 0}{Lt} \frac{(n^2 - n_0^2)}{w} \right] \quad (1)$$

where μ is the dipole moment of the polar solute molecule in debyes,

N is Avogadro's number,

k is Boltzmann's constant in erg deg^{-1} molecule^{-1},

ϵ_0 is the dielectric constant of the pure solvent,

ϵ is the dielectric constant of the solution,

n_0 is the refractive index of the pure solvent,

n is the refractive index of the solution,

M_2 is the molecular weight of the solute,

d_0 is the density of the pure solvent,

w is the weight fraction of the solute.

Apparatus and Chemicals

Wayne Kerr Universal Bridge (B221), dielectric constant cell, Abbé refractometer connected to a thermostat at 25°C, pure benzene, pure nitrobenzene, pure o-, m- and p-dinitrobenzene, calibrated pipettes of 100 ml and 2 ml capacity, six weighing bottles with plastic caps.

(A suitable dielectric constant cell can be constructed as illustrated in Fig. 1. A 150 pF variable capacitor with ceramic insulation is mounted inside a cylindrical glass jar 7 cm in diameter and 11 cm in height—supplied by C. E. Payne and Sons, Ltd. The capacitor is bracketed to the plastic lid and the plates are adjusted by a knob connected to the capacitor with an extension spindle.)

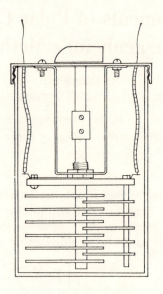

Fig. 1. Dielectric cell.

Method

Solutions of nitrobenzene in benzene containing accurately about 1, 2, 3, 4, 6 and 8 per cent w/w of nitrobenzene are prepared in the weighing bottles. The refractive indices of these solutions and also that of pure benzene are determined at 25°C using the Abbé refractometer. A graph of $(n^2 - n_0^2)/w$ against w is plotted. This should be a horizontal straight line and the intercept gives an average value of the second term in the bracket in equation (1).

The capacitor and container are cleaned and dried and clamped in the thermostat so that the dielectric cell is about three-quarters immersed. The capacitance of the capacitor in the fully opened and fully closed positions is determined with the Wayne Kerr bridge. Two hundred millilitres of pure benzene is pipetted into the cell. From the calibration of the pipette and the density of benzene the weight of benzene added is calculated. When the benzene has attained thermal equilibrium the minimum and maximum capacitances are again determined. Two millilitres of nitrobenzene is pipetted into the benzene, the solution is mixed by opening and closing the capacitor a few times and the minimum and maximum capacitances are again determined. Further additions of nitrobenzene up to about 8 per cent w/w are made from the calibrated pipette which allows the weight fractions to be calculated each time, and the two capacitances are determined.

The dielectric constant for benzene is given by

$$\epsilon_0 = \frac{\Delta C_{\text{benzene}}}{\Delta C_{\text{air}}} \tag{2}$$

where ΔC is the difference between the maximum and minimum capacitances. The dielectric constants of the solutions are calculated in the same manner.

A graph of $(\epsilon - \epsilon_0)/w$ against w is plotted. This should be a straight line and the intercept gives the value of the first term in the bracket in equation (1). The dipole moment of nitrobenzene is then calculated from equation (1).

The experiment is repeated using the three disubstituted isomers.

Compare these results with the dipole moments calculated on the basis of the hexagonal ring structure for benzene.

Discuss the relative importance of the dielectric constant and refractive index terms for each case.

[1] J. W. SMITH, *Electric Dipole Moments*, chap. 2D.
[2] E. A. GUGGENHEIM, *Trans. Far. Soc.*, **45**, 714 (1949).
[3] E. A. GUGGENHEIM, *Trans. Far. Soc.*, **47**, 573 (1951).

Diffusion of Solvent Molecules through High Polymers

R. E. WETTON,

Loughborough University of Technology

Discussion

The diffusion of small organic molecules, such as benzene or ethanol, through a high polymer solid is governed by the rate at which the polymer molecules can rearrange. Free volume theories of diffusion have met with the most success for polymer systems. The fundamental hypothesis of such theories is that a diffusing molecule moves by a series of isolated jumps. A jump will only occur when the free volume fluctuations have created a favourable "hole" for the molecule to enter.

Bueche's[1] ideas on the free volume theory lead to

$$\mathscr{D} = A \exp(-\beta v^{\ddagger}/v_f) \qquad (1)$$

where $\mathscr{D}$ is the intrinsic diffusion coefficient of penetrant through the polymer and $v^{\ddagger}$, the activation volume, represents the critical hole volume required for diffusion. The free volume of the system v_f will be some function of the temperature and volume fraction of the penetrant.

Incorporating the dependence of v_f on the volume fraction of penetrant, V_A, Wilkens and Long[2] obtain the following relation at constant temperature

$$\log \mathscr{D}_T = \log \mathscr{D}_{0T} + \alpha_T V_A \qquad (2)$$

$\mathscr{D}_T$ is the intrinsic diffusion coefficient at volume fraction V_A of penetrant and $\mathscr{D}_{0T}$ is its value as V_A tends to zero. α_T is a constant for a given temperature.

Molecular rearrangement of polymer chains can be considered as a kinetic process and the temperature dependence of the rate is commonly discussed in terms of an Arrhenius type equation. Thus for diffusion coefficients at volume fraction V_A of penetrant the relation is

$$\mathscr{D}(V_A) = \mathscr{D}_{\infty}(V_A) \exp(-\Delta H^{\ddagger}/RT) \qquad (3)$$

Thus, on comparing diffusion data at a temperature T_1 with that at a higher temperature T_2, we have

$$\log\left[\frac{\mathscr{D}_2(V_A)}{\mathscr{D}_1(V_A)}\right] = \frac{\Delta H^{\ddagger}(V_A)}{2 \cdot 303R}\left[\frac{T_2-T_1}{T_2T_1}\right] \tag{4}$$

Hence $\Delta H^{\ddagger}(V_A)$ may be found.

In order to investigate these ideas it is necessary to measure the intrinsic diffusion coefficient ($\mathscr{D}$) as a function of volume fraction of penetrant for at least two different temperatures.

The intrinsic diffusion coefficient is calculated from the mutual diffusion coefficient D from[2]

$$\mathscr{D} = D/(1-V_A)^3 \tag{5}$$

The mutual diffusion coefficient D can only be obtained from experimental observations by approximation methods, because it is an unknown function of penetrant concentration in polymer systems. If, during an experiment, the concentration of penetrant (measured as the volume of penetrant per cubic centimetre of polymer) varies from C_1 to C_2 then an average value of the diffusion coefficient, which is termed $\bar{D}_2$, can be obtained for this concentration range. If this can be repeated from C_2 to C_3 to give $\bar{D}_3$, etc., and the concentration ranges can be made sufficiently small then D can be approximated at any concentration.[3] The average diffusion coefficients $\bar{D}_1, \bar{D}_2, \ldots, \bar{D}_\chi$ are obtained for a series of consecutive concentration intervals ΔC_1, $\Delta C_2, \ldots, \Delta C_\chi$. The sum

$$\sum_{i=1}^{i=\chi} \bar{D}_i \Delta C_i$$

is plotted against C_i and the slope at any concentration is a first approximation to the mutual diffusion coefficient D, at that concentration. The calculation is detailed in the Method section.

Apparatus and Chemicals

The quartz spring apparatus is shown in Fig. 1. This comprises a quartz spring (sensitivity *ca.* 200 cm g^{-1}) mounted in a vacuum system, capable of maintaining 10^{-2} torr. The sample hangs inside a glass tube surrounded by a water thermostat bath. Dewar flasks, cathetometer and stop watch are also required.

Any cross-linked amorphous polymer can be used as a sample provided that it is in film form (a commercial rubber latex for example). Convenient dimensions of the film are $0 \cdot 02 \times 1 \times 4$ cm. The only requirement of the penetrant is that it is thermodynamically compatible with the polymer (i.e. it is a swelling agent) and gives convenient vapour pressures at low temperatures. Benzene, ethanol and hexane are typical convenient penetrants.

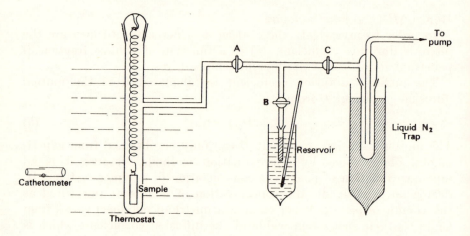

Fig. 1. Quartz spring apparatus.

Method

The average mutual diffusion coefficients $\bar{D}_1$, $\bar{D}_2$, etc., in the concentration ranges 0 to C_1, C_1 to C_2, etc., are obtained by following the sorption kinetics with the quartz spring apparatus shown in Fig. 1. Before the sample is hung on the spring it should be washed in penetrant, dried and measured. The penetrant is placed in the reservoir and air removed by repeated freezing—evacuating—thawing cycles. The penetrant must not be degassed by pumping in the liquid state. After this degassing, the sample is pumped for about 2 hr (with reservoir tap B closed). The reservoir is surrounded by a cold liquid (water) in a Dewar flask. The temperature is chosen to give a suitable low vapour pressure of penetrant.

Diffusion is commenced by opening tap B (with C closed and A open) and simultaneously starting a stop watch. The extension of the spring is recorded as a function of time until no further significant change occurs. Tap B is closed and the reservoir temperature raised a few degrees to give a slightly increased vapour pressure of penetrant. Tap B is opened and extension again recorded against time from the opening. Similar sorption runs are repeated at four successively higher reservoir temperatures, up to room temperature, with the sample temperature constant. From each sorption run, the mass of penetrant absorbed, M_t, at various times is evaluated from the extension. M_∞ is the near equilibrium value at long times. $\bar{D}$ is evaluated from the equation

$$\frac{M_t}{M_\infty} = \frac{4}{l}\left(\frac{\bar{D}t}{\pi}\right)^{1/2} \tag{6}$$

where l is the film thickness. Hence (M_t/M_∞) against $t^{1/2}$ is plotted and the slope determined. Values near equilibrium should not be plotted. From M_∞ the equilibrium mass of the penetrant in the polymer is obtained and assuming that the density of the penetrant in the absorbed state is equal to its liquid value, C_1, C_2, etc., can be obtained in cm³/cm³ of polymer at the end of each sorption run.

Results can now be tabulated as in Table 1.

TABLE 1. TABULATION OF RESULTS

Run	M_∞	C_∞	ΔC	$\bar{D}$	$\bar{D}\Delta C$	$\Sigma\bar{D}\Delta C$
1		C_1	$C_1 - 0$	$\bar{D}_1$	$\bar{D}_1 C_1$	$\bar{D}_1 C_1$
2		C_2	$C_2 - C_1$	$\bar{D}_2$	$\bar{D}_2(C_2 - C_1)$	$\bar{D}_1 C_1 + \bar{D}_2(C_2 - C_1)$
3		C_3	$C_3 - C_2$	$\bar{D}_3$	$\bar{D}_3(C_3 - C_2)$	$\bar{D}_1 C_1 + \bar{D}_2(C_2 - C_1)$ $+ \bar{D}_3(C_3 - C_2)$

$\Sigma\bar{D}\Delta C$ is plotted against the C values and the slope at any concentration is a first approximation to the mutual diffusion coefficient D.

The intrinsic diffusion coefficient $\mathscr{D}$ is then evaluated from equation (5).

The procedure is repeated with the polymer sample held at a different constant temperature and the results are used to investigate the ideas forwarded in the discussion.

1 BUECHE, *Physical Properties of Polymers*, p. 97, Interscience, 1962.
2 WILKENS and LONG, *Trans. Far. Soc.*, **53**, 1146 (1957).
3 CRANK, *The Mathematics of Diffusion*, chap. xi, Oxford University Press, 1956.

Interpretation of X-ray Powder Photographs

D. S. BROWN,

Loughborough University of Technology

Discussion

The crystalline state is characterized by a regular three-dimensional atomic arrangement, which will behave as a three-dimensional diffraction grating to electromagnetic radiation of appropriate wavelength. The effect of a beam of X-rays on a crystal is really a problem in diffraction, but Bragg showed that the theory could be simplified if it were regarded as one of reflection from a series of sets of parallel planes in the crystal. The condition for reflection is then given by

$$\lambda = 2d \sin \theta$$

where λ is the wavelength of the radiation and θ the angle of incidence to a set of parallel planes of perpendicular spacing d.

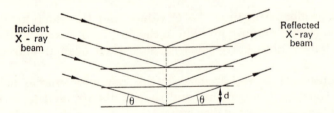

FIG. 1. Set of parallel crystal planes of perpendicular spacing d.

Clearly, for monochromatic radiation, there is only one position where a particular set of planes will reflect X-rays, i.e. there is only one value of θ which satisfies the Bragg equation.

The sets of planes which give rise to the reflections can be defined by the fractional intercepts they make with the three unit cell axes, a, b and c, of the crystal. The reciprocals of these intercepts must always be integers and are defined as the three Miller indices h, k and l. Thus the plane described by the Miller indices 021 ($h = 0$, $k = 2$, $l = 1$) will make intercepts of ∞, $\frac{1}{2}$ and 1 with the a, b and c axes respectively. The d value for this plane will be the perpendicular drawn from the origin to the plane. Clearly, the value of d depends on the unit cell dimensions and the values of h, k and l. High order reflections (high values of h, k and l) will have small d and high θ values, but the number of reflections observable is restricted by the limit of θ at $90°$

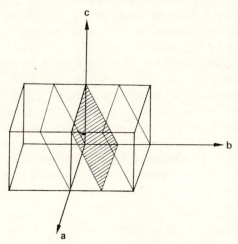

Fig. 2. Orientation of the set of planes (021) relative to the unit cell axes.

For a cubic cell of side a_0, the perpendicular spacing d for any plane of Miller indices h, k, l, is given by

$$d_{hkl} = \frac{a_0}{\sqrt{(h^2 + k^2 + l^2)}}$$

From Bragg's law we have

$$\lambda = 2d \sin \theta$$

hence

$$\sqrt{(h^2 + k^2 + l^2)} = (2a_0/\lambda) \sin \theta$$

or

$$(h^2 + k^2 + l^2) = (4a_0^2/\lambda^2) \sin^2\theta$$

or

$$\log(h^2 + k^2 + l^2) = \log(4a_0^2/\lambda^2) + 2 \log \sin \theta \tag{1}$$

In single crystal X-ray photographs, each spot on a film corresponds to reflection from a given set of planes (hkl) and may be indexed according to the values of h, k and l. In an X-ray photograph of a powder, where all possible orientations of the crystallites are possible, the

X-ray beam is diffracted as a series of cones, one for each set of planes, and these appear as rings on the film. It sometimes happens that different planes have the same Bragg angle θ and this causes reflections from them to superimpose on a powder photograph.

Interpreting X-ray photographs is basically concerned with assigning to each diffracted beam its appropriate Miller indices. In powder photographs, interpretation is relatively simple for cubic crystal systems but less so for other systems. For analytical work, however, a powder photograph provides a "fingerprint" of a crystalline material and unknown samples can be identified by matching the experimental diffraction pattern with a library of known patterns.

In the present investigation, a powder photograph is to be taken of a sample with a cubic crystal system. The lines on the film are to be indexed and the cell dimension and lattice symmetry determined. The substance is to be identified by comparison with known data and in simple cases a trial structure may be proposed.

Apparatus and Chemicals

X-ray generator and tube, CuK_α radiation, X-ray powder camera, small agate mortar and pestle.

Method

A small quantity of a powder having a cubic structure is ground up as finely as possible and placed in a thin-walled Pyrex capillary tube. The tube is mounted in the powder camera and its position adjusted so that it will remain in the X-ray beam during rotation. The camera is exposed to X-rays for an appropriate period, after which the film is developed, fixed and washed in the usual way.

Values of $\log(h^2 + k^2 + l^2)$ are plotted to scale on a strip of graph paper for all possible values of h, k and l up to, say, 4. The last value plotted will therefore be $\log(4^2 + 4^2 + 4^2)$ although in some cases it may be necessary to plot higher values.

Values of θ may be obtained by measuring the diameter, x, of each powder ring across the point on the photograph where the primary X-ray beam leaves the camera. This point can be recognized as being the centre of the more intense arcs. The general relationship between x and θ is $x/r = 4\theta$ radians.

In making measurements for high values of θ, it will be noticed that the lines are resolved into doublets due to the doublet nature of CuK_α radiation. ($CuK_{\alpha 1}$, $\lambda = 1 \cdot 540$ Å; $CuK_{\alpha 2}$, $\lambda = 1 \cdot 544$ Å). If an average value of the wavelength is to be used, measurements can be made to the centre of gravity of each pair of lines.

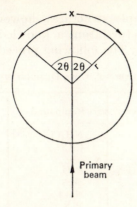

Fɪɢ. 3. Diagrammatic representation of the experimental arrangement.

Values of $2 \log \sin \theta$ are plotted on a second strip of graph paper on the same scale as the $\log(h^2 + k^2 + l^2)$ values. The two strips are placed together and moved relative to one another until coincidence between the two sets of plotted points is obtained. It should be borne in mind that the most accurate measurements will be for high values of θ and hence a better fit is expected for these values.

By comparing the two graphs it is now possible to index each observed line on the powder photograph with its appropriate Miller indices. Systematic absences of certain types of reflections may be noticed and these are related to the lattice symmetry as shown in Table 1.

TABLE 1. LATTICE SYMMETRY RELATIONS

Type of lattice	Conditions for X-ray reflection
Primitive lattice	No consistent absences
Body centred cubic	hkl only observed when $h+k+l = 2n$
Face centered cubic	hkl only observed when $h+k = 2n$ $h+l = 2n$ and $k+l = 2n$

From equation (1) it can be seen that the value of $\log(h^2 + k^2 + l^2)$ is $\log(4a_0^2/\lambda^2)$ when $2 \log \sin \theta = 0$. Hence the cell dimension, a_0, may be calculated assuming that $\lambda = 1.542$ Å for CuK_α radiation.

The specimen is then identified by comparison with the data in Table 2.

The information so far obtained is usually combined with a knowledge of the intensities of the lines, in order to determine the crystal structure. For very simple structures, however, it may be possible to guess at a trial structure from a knowledge of lattice symmetry and cell dimension alone. The following procedure would be adopted.

TABLE 2. CELL DIMENSIONS

Type of lattice	Compound	a_0 Å
Primitive lattice	NH_4Cl	3·87
	CsCl	4·11
Face centred lattice	CaO	4·98
	MgS	5·19
	NaCl	5·63
	PbS	5·92
	NaBr	5·96
	KCl	6·28
	KBr	6·59
	KI	7·05
Body centred lattice	$\beta - ZnB_2O_4$	7·47
	Sc_2O_3	9·84
	Ce_2S_3	8·63

The number of molecules, n, to be arranged in each cell may be determined from the density, ρ, of the crystal, since

$$\rho = \frac{\text{mass of a unit cell}}{\text{volume of a unit cell}} = \frac{nM}{N} \ (a_0^3 \times 10^{-24})$$

where M is the molecular weight of the substance and N is Avogadro's number. The n molecules must be arranged in the unit cell to satisfy the lattice symmetry and in addition, ions must not approach each other closer than their ionic radii allow. With the aid of a model or drawing, a trial structure may be formulated from which accurate bond lengths can be calculated.

Delocalization Energies of Π Molecular Orbital Systems

Discussion

Benzene possesses six π electrons and according to the molecular orbital theory these six π electrons occupy molecular orbitals which extend over all six carbon atoms. These six π electrons are therefore completely delocalized. The bond lengths between the carbon atoms are equal in length (1·39 Å). We represent benzene, as described by molecular orbital theory, by structure I. The energy of this structure is a minimum one and is lower than any other hypothetical structures, for examples II or III.

One possible hypothetical structure is the Kekulé one (II) where the distances between the carbon to carbon centres are not equal. A second hypothetical structure is III where the distances between the carbon to carbon centres are all equal, that is III (and I) are regular hexagons. The energy between I and II is the empirical resonance energy (E.R.E.), whereas the energy between I and III is the vertical resonance energy. Both are positive quantities. The energy between II and III is the distortion energy and according to Coulson and Altmann[1] its value is $+27$ kcal mole^{-1}.

A number of methods are available for the calculation of empirical resonance energies.[2] Despite some disadvantages, the calculation from heats of combustion provides a reasonably accurate method, particularly for aromatic compounds where the resonance energy is large. In order to calculate the heat of combustion of the hypothetical structure, it is necessary to know the energy contributions made by various types of bonds within this structure. These have been estimated by a number of workers,[3] the bond contributions required for the calculation of the resonance energy of benzene are listed in Table 1.

TABLE 1. BOND CONTRIBUTIONS

Bond	Contribution
C—H	54·0
C—C	49·3
C=C	117·4

All quantities are in kcal/mole. A correction of $+1\cdot0$ is required for the presence of the six-membered ring in benzene.

The heat of combustion of the alternating bond representation is calculated from the bond contributions. The difference between this value and the experimental heat of combustion is the empirical resonance energy (E.R.E.).

The energies listed are the contributions made by the various types of bonds to the molecule in the gaseous state. The gaseous state is preferred as it minimizes the problems of molecular interaction which would be more serious in the liquid or solid state. If the heat of combustion determined is for the liquid state, it must be corrected:

$$\Delta H_g = \Delta H_l + \Delta H_v \qquad (1)$$

where ΔH_g is the heat of combustion for the gaseous state, ΔH_l for the liquid state and ΔH_v the heat of vaporization.

If the heat of vaporization is unknown it may be calculated by a convenient method suggested by Klages.[4] According to Trouton's law for an unassociated liquid the heat of vaporization at $t°C$ is given by

$$\Delta H_v(t) = 0\cdot021(273+t)$$

At 25°C the heat of vaporization is given by the relationship:

$$\Delta H_v(25) = \Delta H_v(t) - \Delta C_p(t-25)$$

where ΔC_p is the average value of the increase in molar heat capacities from liquid to vapour phase within the range 25° to $t°C$. Hence

$$\Delta H_v(25) = 5\cdot733 + 25\Delta C_p + (0\cdot021 - \Delta C_p)t$$

and as $\Delta C_p = 0.015$ kcal mole^{-1} °C^{-1} (approximately), then

$$\Delta H_v(25) = (5.4 + 0.036t)\ \text{kcal mole}^{-1}\ {}^{\circ}\text{C}^{-1} \tag{2}$$

This equation is sufficiently accurate for most calculations.

For a solid (heat of combustion ΔH_s) in order to calculate the heat of combustion corresponding to the gaseous state, the heat of fusion (ΔH_f) is also required. This also should be corrected to 25°C, but as the correction is small the heat of fusion at the melting point is sufficient.

In conjugated systems, in addition to the σ skeleton, a molecular orbital consisting of mobile π electrons is present. Each mobile electron is supposed to move under the influence of an individual field due to the attraction of the nuclei, the repulsion of the σ electrons and the repulsion of other mobile π electrons. To define and use an explicit Hamiltonian at this stage would be difficult. Instead, we assume initially an effective Hamiltonian H which is the same for each mobile electron of the system. The wave equation is therefore

$$H\phi = \epsilon\phi \tag{3}$$

where ϕ is the molecular orbital and ϵ is the eigenvalue.

The molecular orbital is formed as a linear combination of atomic orbitals ψ, i.e.

$$\phi_j = c_{j1}\psi_1 + c_{j2}\psi_2 + \ldots c_{jn}\psi_n \tag{4}$$

where c is the coefficient of a particular atomic orbital in the jth molecular orbital. These coefficients and the eigenvalue may be obtained on solving a set of n simultaneous equations

$$c_1(H_{11} - S_{11}\epsilon) + c_2(H_{12} - S_{12}\epsilon) + \ldots + c_n(H_{1n} - S_{1n}\epsilon) = 0$$
$$c_1(H_{21} - S_{21}\epsilon) + c_2(H_{22} - S_{22}\epsilon) + \ldots + c_n(H_{2n} - S_{2n}\epsilon) = 0$$
$$\cdot$$
$$\cdot$$
$$\cdot$$
$$c_1(H_{n1} - S_{n1}\epsilon) + c_2(H_{n2} - S_{n2}\epsilon) + \ldots + c_n(H_{nn} - S_{nn}\epsilon) = 0 \tag{5}$$

In these equations

$$H_{nn} = \int \psi_n H \psi_n d\tau \quad \text{(Coulomb integral)} \tag{6}$$

$$H_{mn} = \int \psi_m H \psi_n d\tau \quad \text{(Resonance integral)} \tag{7}$$

$$S_{nn} = \int \psi_n \psi_n d\tau = 1 \tag{8}$$

$$S_{mn} = \int \psi_m \psi_n d\tau \quad \text{(Overlap integral)} \tag{9}$$

Hückel assumed that the Coulomb integral was significant only in the region of the nth centre and for a system consisting of n identical carbon centres the integrals were equal. The resonance integral was considered to be significant only when m and n were neighbours and again for a system of identical centres the resonance integrals were assumed to be equal. For convenience we write the Coulomb and resonance integrals as α and β respectively. Finally Hückel assumed S_{mn} to be zero.

On adopting these assumptions, the simultaneous equation for a π electron system, e.g. butadiene (IV) can be written simply

$$H_2C=CH-CH=CH_2$$

IV

$$
\begin{array}{llll}
c_1(\alpha-\epsilon)+c_2\beta & +0 & +0 & = 0 \\
c_1\beta & +c_2(\alpha-\epsilon)+c_3\beta & +0 & = 0 \\
0 & +c_2\beta & +c_3(\alpha-\epsilon)+c_4\beta & = 0 \\
0 & +0 & +c_3\beta & +c_4(\alpha-\epsilon) = 0
\end{array}
\tag{10}
$$

A simplified secular determinant can be obtained by putting $x = (\alpha-\epsilon)/\beta$, viz.

$$
\begin{vmatrix}
x & 1 & 0 & 0 \\
1 & x & 1 & 0 \\
0 & 1 & x & 1 \\
0 & 0 & 1 & x
\end{vmatrix}
\tag{11}
$$

On expansion this determinant gives the equation

$$x^4 - 3x^2 + 1 = 0 \tag{12}$$

This can be reduced to a simple quadratic in y on substituting $y = x^2$, viz.

$$y^2 - 3y + 1 = 0 \tag{13}$$

This simple quadratic equation is easily solved and four values for ϵ are obtained.

$$\epsilon_1 = \alpha + 1 \cdot 6$$

$$\epsilon_2 = \alpha + 0 \cdot 6$$

$$\epsilon_3 = \alpha - 0 \cdot 6$$

$$\epsilon_4 = \alpha - 1 \cdot 6 \tag{14}$$

In order to calculate the total π energy E_π, we select the lowest two levels ϵ_1 and ϵ_2 (β is a negative quantity) and place four π electrons in these levels while observing the Pauli exclusion principle

$$E_\pi = 2\epsilon_1 + 2\epsilon_2 \tag{15}$$

or

$$E' = 4\alpha + 4 \cdot 4\beta \tag{16}$$

If the four π electrons had been placed in two localized double bonds as indicated by IV the π energy E'_π would have been

$$E_\pi = 4\alpha + 4\beta \tag{17}$$

The energy difference between E'_π and E_π is the delocalization energy, DE_π, that is

$$DE_\pi = E'_\pi - E_\pi \tag{18}$$
$$= (4\alpha + 4\beta) - (4\alpha + 4 \cdot 4\beta) \tag{19}$$
$$= 0 \cdot 4(-\beta) \tag{20}$$

In this way the delocalization energy is obtained in units of $(-\beta)$ and as β is a negative quantity the delocalization energy is therefore positive.

The delocalization energy is defined in this way by Daudel, Lefebvre and Moser,[5] and by Higasi, Baba and Rembaum.[6] Many other workers write

$$DE_\pi = E_\pi - E'_\pi$$

that is the delocalization energy is obtained in units of $(+\beta)$. Both definitions give the value of the delocalization energy a positive sign, but whereas the first definition gives

$$n(-\beta) = +DE_\pi$$

that is

$$\beta = -DE_\pi/n$$

negative in sign as it should be; the second definition gives

$$n\beta = +DE_\pi$$

that is we have to state that β is a negative quantity and divide across by $-n$ in order to obtain a negative value for β.

Apparatus and Chemicals
Bomb calorimeter and accessories, benzene.

Method
The heat of combustion of benzene is determined. Benzene is a liquid therefore a known weight is sealed off in a thin glass bulb. A known weight of benzoic acid is placed on the bulb. On combustion the benzoic acid breaks the bulb and the benzene is combusted. The heat

released on burning the benzene is the total heat released minus the heat released by the benzoic acid. Other details of the method are available in Experiment 16. The heat of combustion is corrected to the gaseous state using equation (1).

The heat of combustion is calculated from the bond contributions given in Table 1. The empirical resonance energy is the difference between the calculated heat of combustion and the corrected heat of combustion.

Benzene constitutes a system of six π electrons moving in a molecular system of six equal carbon bonds. In the light of the above discussion for butadiene the secular determinant for benzene should be written: (a) in terms of α, β and ϵ; (b) in terms of $x = (\alpha - \epsilon)/\beta$. This determinant on expansion would give a sixth-order equation but if $y = x^2$ is substituted a cubic equation in y is obtained. The value of the roots of this equation may be obtained by factorization. Alternatively, as the determinant is a cyclic one the value of x may be obtained from the formula

$$x = -2\cos(2\pi k/n) \tag{21}$$

where $k = 1, 2, 3 \ldots n$ and $n = 6$. Once the value of the roots have been obtained the delocalization energy DE_π can be calculated as in equation (18) where E_π' is the energy found by placing six π electrons in the three double bonds as indicated in III (assigning $\alpha + 2\beta$ per double bond) and E_π is obtained by placing two π electrons in the three lowest levels ϵ_1, ϵ_2 and ϵ_3. This is the value of DE_π in $(-\beta)$ units.

If this value is equated to the E.R.E. we have

$$DE_\pi = r \, \text{kcal mole}^{-1}$$

and

$$\beta = -r/n \, \text{kcal} \tag{22}$$

Alternatively if the value of DE_π is equated to the V.R.E. we have

$$DE_\pi = t$$
$$= r + s \, \text{kcal mole}^{-1}$$

and

$$\beta = -(r+s)/n \, \text{kcal} \tag{23}$$

Finally the E.R.E. values for other π molecular systems can be obtained from Wheland,[7] and the corresponding DE_π values (in $-\beta$ units) from Higasi *et al.*[6] or Streitwieser[8] (in $+\beta$ units). A plot of E.R.E. and DE_π/β should be prepared and the value of β calculated from the slope (the linear graph is assumed to pass through the origin).

The variations in the value of β obtained by equations (22) and (23) and the graph should be discussed and explained.[9]

[1] C. A. COULSON and S. ALTMANN, *Trans. Faraday Soc.*, **48**, 293 (1952).

[2] E. W. WHELAND, *Resonance in Organic Chemistry*, John Wiley and Chapman & Hall.

[3] E. W. WHELAND, *Ibid.*

[4] F. KLAGES, *Chem. Ber.*, **82**, 358 (1949).

[5] R. DAUDEL, R. LEFEBVRE and C. MOSER, *Quantum Chemistry*, Interscience Inc., 1959.

[6] K. HIGASI, N. BABA and A. REMBAUM, *Quantum Organic Chemistry*, Interscience Inc., 1965, p. 84.

[7] G. W. WHELAND, *Resonance in Organic Chemistry*, J. Wiley, 1955, p. 98.

[8] A. STREITWIESER, Jr., *Molecular Orbital Theory for Organic Chemists*, J. Wiley, 1961.

[9] C. A. COULSON, *Valence*, O.U.P., 1952, p. 235; F. A. COTTON, *Chemical Applications of Group Theory*, Interscience Inc., 1963, p. 287.

K

Thermodynamic Equilibrium Constant of a Molecular Charge Transfer Complex

Discussion

Charge transfer spectra occur when the ground state of a molecule transfers an integral charge (or a major portion of an integral charge) to an excited state. If the charge transfer occurs within a single molecule the process is known as an intramolecular charge transfer. If the charge transfer occurs between adjacent or loosely bonded separate molecules the process is known as an intermolecular charge transfer and the system is referred to as a molecular charge transfer complex. Molecular charge transfer complexes between aromatic hydrocarbons and small inorganic molecules e.g. chlorine, bromine, iodine, interhalogens, oxygen and sulphur dioxide have been widely studied.[1] Aromatic hydrocarbons also form molecular charge transfer complexes with other organic molecules, e.g. tetracyanoethylene (TCNE), I.

I

Mulliken[2] considers that such complexes arise from a Lewis acid–base interaction. According to Lewis, a base is any substance which donates a pair of electrons. An aromatic hydrocarbon such as benzene which is rich in π-electrons could behave in such a way as depicted in II.

II

The corresponding acid would be any substance which accepts the pair of electrons and Merrifield and Phillips[3] immediately on its discovery demonstrated that in TCNE the π-electrons were readily available where an association equilibrium may be written

Base + TCNE ⇌ Molecular charge transfer complex

or, more briefly

$$B + A \rightleftharpoons AB$$

The equilibrium constant for this association is

$$K = \frac{[AB]}{[A][B]} \tag{1}$$

where the concentrations of the acid, base and complex are expressed in mole fraction units and the ratio of the activity coefficients $(\gamma_{AB}/\gamma_A\gamma_B)$ is assumed to be unity.

In the case of an aromatic hydrocarbon such as benzene the molecular charge transfer complex with TCNE gives an intensely coloured solution where the absorbance A' is given by the Beer–Lambert law

$$\beta = \frac{A'}{d} = \frac{1}{d}\log\frac{1}{T} \tag{2}$$

where d is the light path in centimetres, T is the transmittance and β is the product of concentration and molar absorptivity. Thus

$$\beta = \epsilon_A[A] + \epsilon_B[B] + \epsilon_{AB}[AB] \tag{3}$$

where the ϵ_i terms are the molar absorptivities. If the original amounts of A and B are added to the solution are known, then

$$a = [A] + [AB] \tag{4}$$

$$b = [B] + [AB] \tag{5}$$

where a and b are the total amounts of A and B, respectively, which were originally present. Again, if the amount of B is very much in excess of the amount of A, then equation (5) can be rewritten as

$$b \simeq [B] \tag{6}$$

The equilibrium constant can then be expressed as

$$K = \frac{[AB]}{(a - [AB])b} \tag{7}$$

whence

$$[AB] = Kab/(1 + Kb) \tag{8}$$

On substituting equations (4), (6) and (8) into equation (3) we have

$$\beta = \epsilon_A a - \frac{Kab}{1 + Kb} + \epsilon_B b + \epsilon_{AB}\frac{Kab}{1 + Kb} \tag{9}$$

or

$$\beta - \epsilon_A a - \epsilon_B b = \frac{Kab}{1 + Kb} (\epsilon_{AB} - \epsilon_A) \tag{10}$$

Rearranging

$$\frac{a}{\beta - \epsilon_A a - \epsilon_B b} = \frac{(1 + Kb)(\epsilon_{AB} - \epsilon_A)}{Kb} \tag{11}$$

$$= \frac{1}{K(\epsilon_{AB} - \epsilon_A)} \cdot \frac{1}{b} + \frac{1}{(\epsilon_{AB} - \epsilon_A)} \tag{12}$$

In this equation, ϵ_A, ϵ_B and β can be found experimentally and independently by measuring the absorbances of A, B, and the mixture at the same wavelength. Hence, a graph of $a/(\beta - \epsilon_A a - \epsilon_B b)$ against $1/b$ can be plotted. This is a linear graph and is known as the Benesi–Hildebrand graph[4] where the slope is given by $1/K(\epsilon_{AB} - \epsilon_A)$ and the intercept by $1/(\epsilon_{AB} - \epsilon_A)$. Hence we have

$$\frac{\text{intercept}}{\text{slope}} = \frac{1}{(\epsilon_{AB} - \epsilon_A)} \cdot K(\epsilon_{AB} - \epsilon_A) = K \tag{13}$$

We can, therefore, calculate the change in the standard free energy from

$$\Delta G^0 = -RT \ln K \tag{14}$$

or

$$\Delta G^0 = -2 \cdot 303 RT \log K \tag{15}$$

Apparatus and Chemicals

Unicam SP.800 spectrophotometer, three 25 ml graduated flasks, four 10 ml graduated flasks, 1 ml pipette, 3 ml pipette, 5 ml pipette, stock solution of 0·0368 g of TCNE in 1 l. of purified heptane, an aromatic hydrocarbon, e.g. benzene or a series of aromatic hydrocarbons, e.g. benzene, toluene, the xylenes, mesitylene and durene.[5]

The heptane is purified by drying over sodium to remove water, passage through an activated silica gel column to remove other aromatic compounds and finally bubbling nitrogen through the heptane to remove dissolved oxygen.

TCNE reacts with water to give HCN, therefore: (a) it is stored in a flask fitted with an automatic burette; (b) it must not be handled, pipetted, etc.; (c) all residues must be returned to a residue bottle; (d) and the laboratory regulations concerning the POISON HCN must be consulted and fulfilled.

Method

If benzene is selected for the investigation the following solutions should be prepared: (i) 5 ml of benzene + 20 ml of stock TCNE solution; (ii) 3 ml of benzene + 22 ml of stock TCNE solution; (iii) 1 ml of benzene + 24 ml of stock TCNE solution; (iv) 3 ml of benzene + 7 ml of stock TCNE solution; (v) 2 ml of benzene + 8 ml of stock TCNE solution; (vi) 1 ml of benzene + 9 ml of stock TCNE solution; and (vii) 5 ml of benzene + 20 ml of heptane.

The absorption spectra of each of the above solutions and of the stock solution of TCNE should be recorded using heptane in the reference cell. The transmittance values at the wavelength of maximum absorbance of the complex should be determined. The values of β at each dilution can therefore be calculated from equation (2). The values of ϵ_A and ϵ_B at this wavelength can be calculated from the relationship $A' = \epsilon_i cd$ where c is the concentration in mole 1^{-1}. The values of a and b can be obtained from equations (4) and (6). Hence a plot of $a/(\beta - \epsilon_A a - \epsilon_B b)$ against $1/b$ can be made. The slope and intercept of this graph should be determined and hence the equilibrium constant and $\Delta G°$ calculated from equations (13) and (15).

If a series of hydrocarbons is investigated the relationship between (a) the charge transfer frequency and (b) the change in free energy of each complex with the ionization potential of the hydrocarbons can be determined.[5]

[1] L. J. ANDREWS, *Chem. Rev.*, **54**, 713 (1954).

[2] R. S. MULLIKEN, *J. Amer. Chem. Soc.*, **74**, 811 (1952).

[3] R. E. MERRIFIELD and W. D. PHILLIPS, *J. Amer. Chem. Soc.*, **72**, 4677 (1950).

[4] H. A. BENESI and J. H. HILDEBRAND, *J. Amer. Chem. Soc.*, **71**, 2703 (1949.).

[5] W. E. WENTWORTH, G. W. DRAKE, W. HIRSCH and E. CHEN, *J. Chem. Ed.* **41**, 373 (1964).

Thermodynamic Functions for Acid–Base Equilibria

Discussion

The dissociation constant is a function of temperature. If the values of the dissociation constant are available over a suitable range of temperature, it is possible to determine the thermodynamic quantities, ΔH^0 the increase of standard enthalpy, ΔS^0 the increase of standard entropy and ΔC_p^0 the increase of molar heat capacity. The increase in standard free energy ΔG^0 may be determined at a single temperature.

A number of equations have been suggested for the variation of the dissociation constant with temperature.[1] One of the most satisfactory may be written[2]

$$R \ln K = -A/T + C - DT \tag{1}$$

where the parameters A, C and D are determined experimentally.

The increase in standard free energy ΔG^0 at 25°C can be obtained from the equation:

$$\Delta G^0 = -RT \ln K = (A - CT + DT^2) \tag{2}$$

The increase in standard enthalpy ΔH^0 is given by

$$\Delta H^0 = \frac{-T^2 \, d(\Delta G^0/T)}{dT} = RT^2 \, d \ln K/dT = A - DT^2 \tag{3}$$

The increase in standard entropy ΔS^0 may be calculated from the relationships

$$\Delta S^0 = -d\Delta G^0/dT = (C - 2DT) \tag{4}$$

or from

$$\Delta S^0 = \frac{\Delta H^0 - \Delta G^0}{T} \tag{5}$$

The increase in molar heat capacity at constant pressure can be obtained from the relationships

$$\Delta C_p = d\Delta H^0/dT = T \, d\Delta S^0/dT = -T \, d^2\Delta G^0/dT^2 \tag{6}$$

hence

$$\Delta C = -2_p^0 DT \tag{7}$$

The Harned and Owen relationship (1) is an empirical one and a new relationship based on the van't Hoff isochore is available in the literature.[3]

Apparatus and Chemicals

As for Experiment 80, 81 or 82, benzoic acid or a nuclear substituted acid.

Method

The dissociation constant of a weak acid may be determined accurately by one of the following methods:[4]

(a) conductivity method, see Experiment 82;
(b) potentiometric method, see Experiment 80;
(c) spectrophotometric method, see Experiment 81.

The dissociation constant of the acid is determined over the temperature range 15° to 45°C, values being obtained at intervals of 5°.

The values of the parameters A, C and D for equation (1) are calculated.

The thermodynamic quantities ΔG^0, ΔH^0, ΔS^0, and ΔC_p are calculated from equations (2), (3), (4), (5), and (7).

[1] For a discussion of the variation of the dissociation constant with respect to temperature, and the study of thermodynamic functions for acid–base equilibria, reviews are available: D. H. EVERETT and W. F. K. WYNNE-JONES, *Trans. Faraday Soc.*, **35**, 1380 (1939); D. J. C. IVES and J. H. PRYOR, *J. Chem. Soc.*, 2104 (1955); HARNED and OWEN, *Physical Chemistry of Electrolytic Solutions*, Reinhold (1958); R. P. BELL, *The Proton in Chemistry*, Methuen (1959); ROBINSON and STOKES, *Electrolyte Solutions*, Butterworths Scientific Publications (1959); E. J. KING, *Acid–Base Equilibria*, Pergamon Press, Oxford (1965); and J. M. WILSON, N. E. GORE, J. E. SAWBRIDGE and F. CARDENAS-CRUZ, *J. Chem. Soc.*, **B**, 852 (1967).

[2] H. S. HARNED and R. A. ROBINSON, *Trans. Farad. Soc.*, **36**, 973 (1940).

[3] E. C. W. CLARKE and D. N. GLEW, *Trans. Faraday Soc.*, **62**, 539 (1966).

[4] For a comparison of these methods for benzoic acid, see R. A. ROBINSON, Ionization Constants by Spectrophotometry, W. J. HAMER (Ed.), *The Structure of Electrolyte Solutions*, John Wiley.

A Comparison of the Efficiency of Laboratory Fractionating Columns

J. A. SANDBACH,

Wolverhampton College of Technology

Discussion

During a fractional distillation there occurs a continual transfer of the more volatile component from the cooler descending liquid to the warmer ascending vapour. Simultaneously a transfer in the opposite direction of the less volatile component takes place. By this means a separation is achieved. The magnitude of the separation will depend on the efficiency of the column and the pair of liquids involved.

In this experiment the efficiencies of different columns are compared by observing the compositions at the top and bottom of the column when the still is charged with a standard pair of liquids (toluene and methylcyclohexane) and a steady state has been reached.

Column efficiencies are expressed in terms of the number of theoretical plates to which they are equivalent. The concept of a theoretical plate can be understood by referring to Fig. 1. A liquid of composition (a) will be in equilibrium with its vapour (b). If the vapour is removed and completely condensed, the liquid formed will also have the composition (b). A theoretical plate is a device capable of achieving just such a separation. The number of theoretical plates to which a column is equivalent is then equal to the number of "steps" on Fig. 1 separating the compositions of the liquid at the bottom of the column and the condensed vapour at the top. In practice, it is the liquid in the still and the liquid refluxing above the top of the column which are analysed. Since a separation roughly equivalent to a theoretical plate occurs in the still pot itself the column efficiency is one less than the number of steps on Fig. 1 between the samples.

Apparatus and Chemicals

Abbé refractometer, fractionating columns (a lagged column 40 cm long and 2 cm in diameter packed with Fenske helices and a similar Vigreux column will be found suitable), 250 ml two-necked flask, a

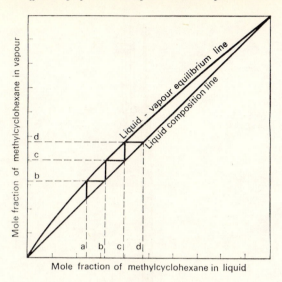

Fig. 1. Liquid–vapour equilibria of mixtures of methylcyclohexane and toluene.

modified Claisen head, thermometer, reflux condenser, heating mantle, semi-micro pipettes, measuring cylinders, pure dry toluene and pure dry methylcyclohexane.

Method

The distillation apparatus is assembled as shown in Fig. 2. Toluene (90 ml) and methylcyclohexane (10 ml) are placed in the still and the mixture is boiled vigorously until the packed column is completely flooded. The heat is reduced until a gentle reflux is taking place above the top of the column. After a steady state has been maintained for about half an hour, three small samples are removed from the top of the column. The first two are rejected. The composition of the third and of a sample taken from the still pot are determined by measuring their refractive indices. The procedure is the same for the Vigreux column except that preflooding is not necessary and a steady state is established more rapidly.

From the data given below, plots of refractive index against composition (as mole fraction of methylcyclohexane) and vapour composition against liquid composition (Fig. 1) are drawn. The number of theoretical plates for each column is determined and the height equivalent to a theoretical plate (H.E.T.P.) is calculated by using the relation

$$\text{H.E.T.P.} = \frac{\text{number of theoretical plates}}{\text{column height}}$$

к*

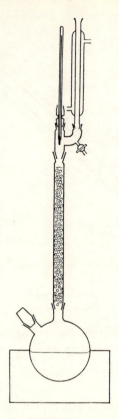

FIG. 2. Distillation apparatus.

TABLE 1. DATA FROM MIXTURES OF TOLUENE AND METHYLCYCLOHEXANE[1]

Liquid composition (mole fraction methylcyclohexane)	Vapour composition	$n_D^{20^0}$ (liquid)
0·000	0·000	1·4965
0·100	0·143	1·4871
0·200	0·270	1·4782
0·300	0·378	1·4699
0·400	0·470	1·4620
0·500	0·560	1·4544
0·600	0·650	1·4474
0·700	0·737	1·4408
0·800	0·818	1·4345
0·900	0·906	1·4286
1·000	1·000	1·4235

[1] *Technique of Organic Chemistry*, Interscience Publishers Inc., vol. iv, p. 71.

The investigation may be extended to other types of packing, for example, Dixon gauze rings and glass beads would be suitable. The effect of replacing the lagging of the packed columns by heating tape or of dispensing with some of the lagging could usefully be studied. Alternatively, the column efficiency could be measured when distillate is steadily removed and the result compared with the efficiency under total reflux.

The Angle of Twist around an Essential C — C Single Bond for a Series of Substituted Carbonyl Compounds

Discussion

If light from a suitable source is passed through a transparent medium before passing through a prism the medium abstracts certain wavelengths from the spectrum of the source, and an absorption spectrum is produced. This spectrum can be studied by photographing it. It is more usual to measure the fraction of light absorbed by a photoelectric cell, and plot the absorption (or the transmission) against the wavelength (or wavenumber). Strictly speaking, a plot of this type is an absorption curve. Absorption spectra data are usually presented in this way.

Absorption intensity is expressed in terms of absorptivities. For a solution, the molar absorptivity at a given wavelength is related to the absorbance thus:

$$\epsilon = A/cl \tag{1}$$

where c is the concentration (mole l^{-1}) and l is the width of the cell in centimetres. The absorption curve is drawn with molar absorptivity ϵ as ordinate, and wavelength λ as abscissa. The absorption maximum $(= \lambda_{max})$ is the wavelength at which a peak $(= \epsilon_{max})$ occurs in the absorption curve. These wavelengths of the absorption maxima provide a means of identifying a functional group and hence a particular compound.

When light is absorbed, the energy of the molecule is momentarily increased by an amount equal to that of a photon. The energy absorbed is related to the frequency of the light by the equation

$$\Delta E = h\nu$$

where E is the energy, h is Planck's constant, and ν the frequency. In general, for rotational and vibrational changes within a molecule, small energies of 0·1 to 10 kcal per mole are required. These energy changes may be brought about by absorption in the infra-red. In contrast, electronic transitions from a lower energy level to a higher energy level, require larger increments of energy of the order of 30 to 300 kcal per mole. For such energy changes visible, and the even more powerful ultra-violet, radiation may be required.

In general the energy levels are shown as in Fig. 1. In the bonding levels π electrons have higher energies than the corresponding σ electrons. In the antibonding levels this order is reversed. Finally the non-bonding lone pair electrons are less strongly bound in the molecule than the bonding electrons.

Level.

Antibonding	σ^*
Antibonding	π^*
E Non-bonding	n
Bonding	π
Bonding	σ

FIG. 1. Energy levels.

Hence bands arising from $\pi \to \pi^*$ or $\sigma \to \sigma^*$ transitions usually occur in the far ultra-violet region whereas $n \to \pi^*$ transitions frequently occur at longer wavelengths.

Transitions between these levels are possible in the ketone molecule (I)

$$\begin{matrix} R & \pi \\ & \downarrow \\ \sigma \diagdown C = 0 : \leftarrow n \\ R \diagup \end{matrix}$$

I

For example, acetone exhibits three bands at 150 mμ, 190 mμ, and 280 mμ corresponding to $\pi \to \pi^*$, $n \to \sigma^*$ and $n \to \pi^*$ transitions respectively.

In aromatic carbonyl compounds, if conjugation between the carbonyl group and the benzene nucleus is possible, structures III and IV are possible and these bands are modified.

| II | III | IV |

The general effect of conjugation is to displace the bands to longer wavelengths. In acetophenone there is an intense band in the 240 mμ region due to the $\pi \to \pi^*$ transition, a less intense band at 325 mμ due to an $n \to \pi^*$ transition. The presence of transitions, other than $\pi \to \pi^*$ transitions, may be obscured due to transitions within the benzene nucleus or the bands may merge into the $\pi \to \pi^*$ bands when the systems are examined in polar solvents.[1]

The spectra of an aromatic organic molecule is modified when various atomic groups are substituted for the hydrogen atoms attached to the parent substance. The region of selective absorption may be displaced and there may be a change in intensity. Wavelength displacement to longer wavelength regions are called bathochromic shifts, displacements to shorter wavelength regions are called hypsochromic shifts. For a conjugated system where only one bond is twisted the wavelength displacements depend upon the sign and magnitude of the change in the double bond character of the twisted bond on excitation.[2] If the twisted bond has a larger bond order in the ground state than in the excited state a bathochromic shift occurs. A hypsochromic shift occurs if the reverse is true. Changes in intensity are referred to as hyperchromic (an increase in intensity) and hypochromic (a decrease in intensity) effects.

The introduction of alkyl groups into acetophenone gives rise to these effects. An alkyl group in the para position gives rise to a small bathochromic displacement and a hyperchromic effect ($\Delta_\epsilon = +1500$ ca). The introduction of an ortho group may cause a marked hypochromic effect in ketones but not in aldehydes. This hypochromic effect is due to the steric effect of the *ortho*-substituents which restrict rotation about the C—C linkage; this occurs with the ketones but not with the aldehydes where the formyl group replaces the larger acetyl group.

A measure of this restricted rotation is given by the ratio r of the observed values of the molar absorptivity to the anticipated value if steric hindrance had not occurred.[3]

$$r = \frac{\epsilon_{max} \text{ (sterically hindered compound)}}{\epsilon_{max} \text{ (unhindered compound)}} \quad (2)$$

To calculate the value for the unhindered ketone, an increment of $\Delta_\epsilon = 1500$ is added to the molar absorptivity of the parent carbonyl compound for each alkyl substituent in the ortho- or para-positions.

In an ortho-substituted sterically hindered ketone the mean (equilibrium) angle between the planes of the phenyl and the carbonyl group (θ) will be related to r according to Braude[3] by

$$r = \cos^2\theta \tag{3}$$

from which the mean interplanar angles for the ketones under investigation may be calculated.

The interplanar angle can be calculated from the relationship

$$\cos^2\theta = (\mu_{obs} - \mu_{90})/(\mu_v - \mu_{90}) \tag{4}$$

where μ_{obs} is the experimental dipole moment, μ_v is the dipole moment calculated by vector addition (Experiment 54) and μ_{90} is the moment of a completely deconjugated system in which the planes of the phenyl and the carbonyl group are at right angles and which can be taken to be the same as that for a saturated ketone (cf. acetone, $\mu = 2{\cdot}75$).[4]

Apparatus and Chemicals

Ultra-violet spectrophotometer, e.g. Unicam S.P.800, solutions of benzaldehyde, acetophenone, p-methyl acetophenone, o-methyl acetophenone, 2 : 4 : 6-trimethyl acetophenone, absolute alcohol.

Method

Absolute alcohol may be used as a solvent provided it has not been dried by azeotropic distillation. The azeotropic elimination of water by means of benzene renders the solvent useless in the region 200–300 mμ. Traces of benzene may be removed by shaking the alcohol with clean silica-gel. This alcohol may also contain traces of aldehyde. These may be removed by refluxing with 1 per cent sodium hydroxide for 2 to 3 hr followed by fractional distillation.

A $0{\cdot}0001$ M solution of acetophenone is prepared by dilution. The absorbance of the solution is determined from 320 mμ to 220 mμ. The intervals between the observations should be 5 mμ, except in the region of maximum intensity where readings should be made every 1 mμ.

The molar absorptivity is determined using equation (1). A graph is drawn for ϵ against λ. The values of ϵ_{max} are noted in the 240 mμ region.

The experiment is repeated with a solution of the other carbonyl compounds. The value of θ is calculated using equation (3) and equation (4).

Benzaldehyde or an ortho-substituted benzaldehyde should be investigated. Other dimethyl, trimethyl or tetramethyl substituted acetophenones may also be investigated. In all cases a Catalin model of the molecule should be constructed and the effect of the ortho-substituents should be discussed in the light of the model and the experimental results. The interplanar angles derived from spectral and dipole moment data should also be compared.

[1] S. F. MASON, Molecular Electronic Spectra, *Quart. Rev.*, IV, 287 (1961).
[2] C. A. COULSON, *Steric Effects in Conjugated Systems* (ed. G. W. GRAY), Butterworths, S. (1958). HEILBONNER and GERDIL, *Helv. Chim. Acta*, **39**, 1996 (1956).
[3] E. A. BRAUDE and F. SONDHEIMER, *J. Chem. Soc.*, 3754 (1955).
[4] KADESCH and WELLER, *J. Amer. Chem. Soc.*, **63**, 1310 (1941).

Relative Strengths of Hydrogen Bonds by a Spectrophotometric Method

Discussion

A hydrogen bond may be formed when a hydrogen atom of a molecule lies between two electronegative atoms such as oxygen. This can occur with carboxylic acids in inactive organic solvents where due to the geometrical arrangement of the carboxylic groups two hydrogen bonds are preferentially formed giving rise to a dimer I.

$$
\begin{array}{ccc}
 & O \ldots H{-}O & \\
 & \diagup \qquad \diagdown & \\
R{-}C & & C{-}R \\
 & \diagdown \qquad \diagup & \\
 & O{-}H \ldots O & \\
 & I &
\end{array}
$$

These bonds are essentially weak bonds (0–10 kcal/mole^{-1}) compared with the normal O—H bond ($ca.$ 110 kcal) and for each hydrogen bond in the acid dimer $-\Delta H$ has the value 4–8 kcal/mole.

The nature of this bond has been the subject of considerable controversy but in general it would appear that the following energies are involved:[1]

    (a) delocalization energy,
    (b) electrostatic energy,
    (c) dispersion energy, and
    (d) repulsive energies.

Of these factors the delocalization energy is possibly the largest numerically and it is the purpose of this investigation to study the variation in delocalization energy due to inhibition of delocalization by substitution in the ortho positions of the benzene ring.

Benzoic acid in an inactive solvent dimerizes to give a dimer II which can assume a coplanar configuration III,

$$
\begin{array}{ccc}
 & O \ldots H{-}O & \\
 & \diagup \qquad \diagdown & \\
C_1{-}C_2 & & C_2{-}C_1 \\
 & \diagdown \qquad \diagup & \\
 & O{-}H \ldots O & \\
 & II &
\end{array}
$$

$$+ \left\langle \bigcirc \right\rangle C_1 {=} C_2 \underset{O-H \ldots O^-}{\overset{O^- \ldots H-O}{\left\langle \right.}} C_2 {-} C_1 \left\langle \bigcirc \right\rangle +$$

III

A planar configuration is impossible with a bulky methyl group in any of the ortho positions which restricts rotation around the C_1—C_2 bonds and as a result the delocalization energy is reduced.

Benzoic acid shows three absorption bands in the region 200–300 mμ.[2] These three bands can be simply referred to as A, B and C with λ_{max} occurring near 202 mμ, 230 mμ and 275 mμ respectively. Of these we are interested in B and C and both would appear to be associated with benzenoid absorption displacements due to the carboxyl group in the aromatic ring (230 and 274 mμ). Quite simply these bonds may be regarded as due to π to π^* transitions: an intense π to π^* band at 230 mμ ($\epsilon \sim 13,000$) and a forbidden π to π^* band at 275 mμ ($\epsilon \sim 1000$). Finally it should be noted that the introduction of a methyl substituent in the ortho position causes a displacement of the B and C bands to longer wavelengths, and with the introduction of two ortho substituents in the same benzene ring one of these bands (the B band) is eliminated.

Forbes, Knight and Coffen[3] suggested a method of estimating the strengths of hydrogen bonds of substituted benzoic acids based on the observed concentration dependence of the ultraviolet absorption spectra of the acids with the nature and position of the substituent. These workers found that in plots of ϵ_{max} (B band) vs. log c (concentration), a critical value (see Fig. 1) for ϵ_{max} always occurs and this critical concentration they regarded as a measure of the strength of the hydrogen bonds. In sterically hindered acids where hydrogen bonding is weakest a greater concentration of the acid is required to give the critical concentration. Again, as the acids are sterically hindered λ_{max} for the C band is selected (rather than λ_{max} for the B band) and these maximum values for each solution are plotted against log c.

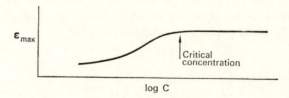

Fɪɢ. 1. Plot of ϵ_{max} vs. log c.

Apparatus and Chemicals

Hilger and Watts Uvispek spectrophotometer and/or Unicam S.P.800 spectrophotometer; pure cyclohexane (fractionation and passage through a silica gel column); 5×50 ml glass-stoppered graduated flasks; benzoic acid, ortho toluic acid, and 2,4,6-trimethylbenzoic acid.

Method

Solutions of the acid in cyclohexane are prepared having accurately known concentrations (c) of about 5×10^{-3}, $2 \cdot 5 \times 10^{-3}$, $1 \cdot 0 \times 10^{-3}$, 1×10^{-4}, and 5×10^{-5} M.

The ultraviolet spectra of each solution is obtained with the spectrophotometer and the value of ϵ_{max} for the C band (275–285 mμ region) is noted. This maximum value for the molar absorptivity of each solution is plotted against the $\log c$ for each solution and the critical concentration assessed.

The critical concentrations indicate the relative strengths of the hydrogen bonds for benzoic acid and its mono- and dimethyl ortho-derivatives. Other derivatives of benzoic acid may be investigated in a similar way and the relative strengths of the hydrogen bonds should be discussed in relation to the influence of the substituents on the delocalization of the acids.

[1] COULSON, *The Hydrogen Bonds Hydrogen Bonding*, HADZI and THOMPSON (Eds.), Pergamon Press (1959); J. COULSON, *Loughborough College's Chem. Soc.* **3**, 38 (1959).

[2] MOSER and KOHLENBERG, *J.C.S.*, 1951, 804.

[3] FORBES, KNIGHT and COFFEN, *Can. J. Chem.* **38**, 728 (1960).

The Evaluation of the Bond Angle, Force Constants and Heat Capacity of Sulphur Dioxide from its Vibrational Spectrum

A. G. BRIGGS,

Loughborough University of Technology

Discussion

Sulphur dioxide is a symmetrical non-linear molecule and has three fundamental modes of vibration (ν_1, ν_2, ν_3) which give rise to infra-red absorption bands. Some of the overtone ($2\nu_i$) and combination ($\nu_i + \nu_j$) bands can also be observed in the infra-red region. The valence force model of sulphur dioxide assumes

(i) that along the direction of each S—O bond there is a restoring force which acts during the period of each vibration and opposes the displacement of the two atoms from their equilibrium internuclear distance;

(ii) that there is a restoring force opposing the change in the angle between the two S—O bonds during a bending vibration.

Use of the above model means that there are more normal vibrational modes than there are force constants so that the latter are overdetermined. This would permit an extra check on the validity of the model if the bond angle were known from another source.

If it be assumed that the normal vibrations of sulphur dioxide are of simple harmonic type then it can be shown that

$$4\pi^2\nu_3^2 = (1 + (2m_O/m_S)\sin^2\alpha)k_1/m_O \tag{1}$$

$$4\pi^2(\nu_1^2 + \nu_2^2) = (1 + (2m_O/m_S)\cos^2\alpha)k_1/m_O$$
$$+ 2/m_O(1 + (2m_O/m_S)\sin^2\alpha)k_\delta/l^2 \tag{2}$$

$$16\pi^4\nu_1^2\nu_2^2 = 2(1 + 2m_O/m_S)(k_1/m_O^2)(k_\delta/l^2) \tag{3}$$

in which m_O and m_S are the masses of single atoms of oxygen and sulphur, respectively; 2α is the angle between the S—O bonds; k_1 is the stretching motion force constant; k_δ/l^2 is the bending motion force constant (l is the equilibrium S—O distance; δ is the change in bond angle during a vibration).

It can further be shown from the above equations that

$$w^3 - (1 + 2m_O/m_S)(v_1^2 + v_2^2 + v_3^2)v_3^2 w / v_1^2 v_2^2 + 2(1 + m_O/m_S)$$
$$(1 + 2m_O/m_S)(v_3^4/v_1^2 v_2^2) = 0 \qquad (4)$$

where $w = 1 + (2m_O/m_S)\sin^2\alpha$.

If v_i is expressed in cm^{-1} units it must be converted to a true frequency (cycles sec^{-1}) by multiplying by the speed of light, $c = 3 \times 10^{10}$ cm sec^{-1}.

The heat capacity at constant volume, C_v, is composed of molecular vibrational, rotational and translational contributions, the last two of these being equal to $(3/2)R$, independent of temperature. The vibrational contribution to C_v, however, is temperature dependent and can be calculated from the vibrational partition function Q_{vib}, provided that the normal modes of vibration are known.

$$Q_{\text{vib}} = \prod_i (Q_i)$$

and, assuming harmonic oscillations,

$$Q_i = \exp(-hv_i/2kT)/(1 - \exp(-hv_i/kT))$$

It can be shown that

$$C_v(\text{vib}) = \partial/\partial T.(RT^2(\partial \ln Q_{\text{vib}}/\partial T))$$

and, substituting for Q_{vib},

$$C_v(\text{vib}) = R \sum_i (u_i^2 \exp(-u_i))/(1 - \exp(-u_i))^2 \qquad (5)$$

where $u_i = hv_i/kT$ and the summation is made over all the normal vibrational modes.

Apparatus and Chemicals

Infra-red spectrometer, e.g. Unicam S.P.200 or Perkin–Elmer Infracord, 10 cm gas cell with sodium chloride windows, cylinder of sulphur dioxide gas with fine control valve, simple vacuum line for evacuating the cell and introducing known pressures of sulphur dioxide.

Method

All spectra should be recorded on the same chart, the base line being displaced downwards a small amount each time. Taps on the gas cell and the vacuum line should be turned slowly to avoid sudden pressure changes. Advice should be obtained from a demonstrator before using the apparatus.

The gas cell and sulphur dioxide cylinder are attached to the vacuum line, all taps being closed. Tap A is opened, then taps B, D and C. After pumping out for a few minutes, A is closed and a check is made that there are no leaks—the mercury level should remain constant. Tap A is opened and pumping continued for 10 min. Taps D and C are closed, then E is opened and the evacuated gas cell is detached.

(*Note*. The cell has a glass body but sodium chloride windows. When not in use it should always be kept in a desiccator. The windows should not be touched or breathed on since fogging of the surface will result. A cold cell is not exposed to a warm atmosphere for the same reason.)

The infra-red spectrum of the evacuated cell is obtained with the base line set at about 90 per cent transmission for frequencies greater than about 800 cm^{-1}. Below 800 cm^{-1} transmission decreases rapidly because of absorption by the cell windows.

The cell is replaced on the vacuum line, E is closed and D and C are opened. After a few minutes A is closed and the sulphur dioxide cylinder valve is very cautiously opened to a minute extent. (*Note*. Before connecting the cylinder to the line it should be checked with a demonstrator that the valve can be opened by hand to the required extent.) Sulphur dioxide gas is allowed to enter the whole system to a pressure of 10–15 torr. Tap C is closed and then the valve on the cylinder, in this order. A is opened extremely slowly and the apparatus is evacuated. D is closed, E is opened, the cell is detached and the low pressure sulphur dioxide spectrum is recorded.

The cell is replaced on the vacuum line. E is closed, D is opened then C is opened very slowly and the cell is evacuated. A is closed and sulphur dioxide gas is slowly admitted to the cell to a pressure of about 730 torr. C is closed and the supply of sulphur dioxide gas is cut off. The apparatus is evacuated and the cell is detached as described above. The high-pressure sulphur dioxide spectrum is recorded.

Finally, the cell is connected to the vacuum system and evacuated for about 10–15 min. D is closed, E is opened very slowly and the cell is returned with its taps open to a desiccator.

It is essential that the wavelength calibration of the spectrometer be checked by the use of reference substances. Polystyrene and a thin film of liquid indene are suitable and the mean of the two calibrations may be used.

In the spectrum of the evacuated cell there will be observed three low-intensity negative peaks. Their spectral positions should be recorded and their origin explained.

The corrected frequencies of the low- and high-pressure sulphur dioxide bands are recorded. Since rotational structure will not be resolved, the frequency of a band origin can be taken as that frequency at which minimum absorption occurs within the band. The appropriate symbols (ν_i, $2\nu_i$, or $\nu_j + \nu_j$) are assigned to the bands using the following information. Bands arising from the fundamental (normal) modes of sulphur dioxide have a much greater intensity than overtone and combination bands and are the only ones to appear in the low pressure spectrum. In general, a symmetrical stretching frequency (ν_1) is lower than the corresponding antisymmetrical one (ν_3). A bending frequency (ν_2) is much lower and for sulphur dioxide occurs beyond the low frequency limit of the instrument, in the potassium bromide region. Thus ν_1 and ν_3 can immediately be assigned to peaks. A study of the high pressure spectrum will enable overtone and combination assignments to be made. The value of ν_2 must be deduced from a combination band and then explained.

The O—S—O bond angle is calculated from equation (4), k_1 is calculated from equation (1) and k_δ / l^2 is calculated from equation (3). Comment should be made on the relative magnitudes of k_1 and k_δ / l^2. k_1 is compared with the force constant for the SO radical, calculated from the data of Norrish and Oldershaw.[1]

From the values of ν_1, ν_2 and ν_3, $C_v(\text{vib})$ is calculated[2] for 298°K and 500°K and C_v at each temperature is deduced. These statistical

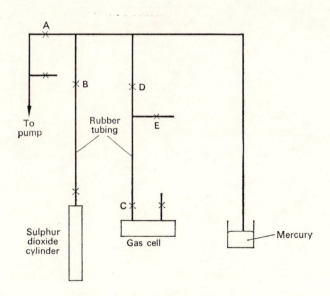

Fɪɢ. 1. Diagram of the apparatus.

thermodynamic values are compared with the directly observed values $C_v = C_p - R = 7 \cdot 3$ cal deg^{-1} mole^{-1} at 298°K and $C_v = 9 \cdot 0$ cal deg^{-1} mole^{-1} at 500°K.

[1] R. G. W. NORRISH and G. A. OLDERSHAW, *Proc. Roy. Soc.*, A **249**, 498 (1958). Full details for the SO radical are given but to evaluate k for SO, the following data will suffice:

$$\omega_e = 1148 \cdot 19 \text{ cm}^{-1}; \quad \omega_e x_e = 6 \cdot 116 \text{ cm}^{-1}.$$

[2] For tabulations of u_i and C_v(vib) see W. J. MOORE, *Physical Chemistry*, 4th ed., p. 632, Longmans, 1963.

The Internuclear Distance of Hydrogen Chloride from its Vibration–Rotation Spectrum

A. G. BRIGGS,

Loughborough University of Technology

Discussion

The purely vibrational term values, in cm^{-1}, for an anharmonic oscillator are given by

$$G(v) = \omega_e(v+\tfrac{1}{2}) - w_e x_e(v+\tfrac{1}{2})^2 + \ldots \tag{1}$$

where $G(v)$ is the energy of the vibrational level with quantum number v, relative to the minimum of the ground electronic state potential energy curve. $\omega_e(cm^{-1})$ is a constant, often referred to as the ground state vibrational frequency and $\omega_e x_e$ (cm^{-1}) is an anharmonicity constant.

For the above vibrator, the purely rotational term values (cm^{-1}) are, neglecting centrifugal distortion,

$$F_v(J) = B_v J(J+1) \tag{2}$$

where $F_v(J)$ is the energy of a rotational level relative to the energy of the non-rotating molecule $(J = 0)$ in the vibrational level v; J is the rotational quantum number; $B_v(cm^{-1})$ is the rotational constant of the molecule.

$$B_v = h/8\pi^2\mu r_v^2 c \tag{3}$$

where h is Planck's constant, r_v is the internuclear distance in centimetres and c is the velocity of light. μ is the reduced mass of the molecule and in the case of hydrogen chloride is defined by

$$1/\mu = 1/m_H + 1/m_{Cl}{}^{35}$$

where m is the atomic mass in grams.

B_v varies with the vibrational quantum number v, the relationship between them being

$$B_v = B_e - \alpha_e(v+\tfrac{1}{2}) \tag{4}$$

B_e is the value which B_v would attain if the molecule were in the (hypothetical) non-vibrating state with the atoms a distance of r_e cm from each other, and α_e is a constant.

Consider the hypothetical, purely vibrational fundamental transitions which will appear at the infra-red frequency given by

$$\nu_0 = G(v') - G(v'') \tag{5}$$

ν_0 (in cm^{-1} units) is called the band origin; v' and v'' refer to the upper and lower vibrational levels, respectively. The upper level has $v = 1$ and the lower level has $v = 0$. For most molecules, every vibrational transition is accompanied by rotational energy changes which give rise to characteristic fine structure.

From equations (1), (2) and (5), the overall structure of a vibration–rotation band can be represented by

$$\nu = \nu_0 + F_{v'}(J') - F_{v''}(J'')$$
$$= \nu_0 + B_{v'}J'(J'+1) - B_{v''}J''(J''+1) \tag{6}$$

where $\nu(cm^{-1})$ is the frequency of a given fine structure line of the band. Single and double primes relate to upper and lower states respectively. In the $v = 0$ level, before absorption of radiation occurs, a number of rotational levels will be populated and transitions will take place to rotational levels in the $v = 1$ state in accordance with the selection rule $\Delta J = \pm 1$. When $\Delta J = +1$, the transition gives rise to an R line and when $\Delta J = -1$ a P line is obtained.

Equation (6) may be simplified by replacing J'' by J. For R and P lines, J' then becomes $(J+1)$ and $(J-1)$ respectively. For the R series of lines, originating from the various populated rotational levels of $v = 0$, equation (6) becomes

$$\nu_R = \nu_0 + B_{v'}(J+1)(J+2) - B_{v''}J(J+1)$$
$$= \nu_0 + 2B_{v'} + (3B_{v'} - B_{v''})J + (B_{v'} - B_{v''})J^2 \tag{7}$$

in which J can take the values $0, 1, 2, \ldots$. Similarly, for the P lines

$$\nu_P = \nu_0 - (B_{v'} + B_{v''})J + (B_{v'} - B_{v''})J^2 \tag{8}$$

in which J can take the values $1, 2, 3, \ldots$.

Equations (7) and (8) can be represented by the single formula

$$\nu = \nu_0 + (B_{v'} + B_{v''})m + (B_{v'} - B_{v''})m^2 \tag{9}$$

where m is an integral running number. For the R branch of a band

$$m = (J+1); \quad m = 1, 2, 3, \ldots$$

and for the P branch

$$m = -J; \quad m = -1, -2, -3, \ldots$$

The R and P branches form a continuous series of lines with one line missing at $m = 0$. The position of this missing line corresponds to the band origin, ν_0.

By subtracting (8) from (7) and simplifying

$$R(J) - P(J) = 4B_{v'}(J + \tfrac{1}{2}) \tag{10}$$

Similarly, if J is replaced by $(J - 1)$ in equation (7) and by $(J + 1)$ in equation (8), then

$$R(J - 1) - P(J + 1) = 4B_{v''}(J + \tfrac{1}{2}) \tag{11}$$

Substitution of J by $(J - 1)$ in equation (7), followed by the addition of equation (7) to equation (8) leads to

$$R(J - 1) + P(J) = 2\nu_0 - 2(B_{v''} - B_{v'})J^2 \tag{12}$$

Hence $B_{v'}$, $B_{v''}$, and ν_0 can be obtained from equations (10), (11) and (12), respectively, by a graphical method.

Transitions for which $\Delta v = 2, 3, \ldots$ are found to occur and give rise to overtone bands. These have only about 1 per cent (or less) of the intensity of the fundamental band, although they exhibit the same general fine structure which may be evaluated as described above.

Hydrogen chloride contains both HCl^{35} and HCl^{37} molecules which will vibrate with slightly different energies because of their different reduced masses. If vibrational anharmonicity is neglected, then the vibrational frequency (cm^{-1}) of the non-rotating molecule is given by

$$\nu = (1/2\pi c)(k/\mu)^{1/2} \tag{13}$$

Here k is the vibrational force constant in dyne cm^{-1} and is assumed to be independent of μ. From equation (13) it can be seen that each HCl^{35} fine structure line in a band will be accompanied by the corresponding line of HCl^{37} which will occur at a slightly lower frequency.[1]

Apparatus and Chemicals

Infra-red spectrometer capable of resolving lines 15 cm^{-1} apart in the region of 3000 cm^{-1}, 10 cm gas cell with sodium chloride windows, source of hydrogen chloride gas, source of hydrogen bromide gas, cylinder of methane.

Method

The gas cell is thoroughly purged out with hydrogen chloride and the infra-red spectrum of the latter is obtained. The spectra of methane and hydrogen bromide are then separately obtained in the same manner and are used to calibrate[2] the wavenumber scale of the instrument. (If the above procedure gives too high a pressure in the gas cell for adequate operation of the spectrometer the cell may be filled to a lower pressure with the aid of a simple vacuum line).

The wavenumber (*in vacuo*) of each line in the HCl^{35} fundamental band, lying between 2640 and 3070 cm^{-1}, is measured. The position of the band origin is located by inspection. The first line on its high frequency side is designated $R(0)$, and on its low-frequency side $P(1)$. The other R and P lines are then numbered in sequence, away from the band origin.

(a) $[R(J-1)-P(J+1)]$ cm^{-1} is plotted against J. B_0 is obtained from the slope of the graph ($= 4B_0$).

(b) $[R(J)-P(J)]$ cm^{-1} is plotted against J. B_1 is obtained from the slope of the graph ($= 4B_1$).

(c) $[R(J-1)+P(J)]$ cm^{-1} is plotted against J^2 and ν_0 is obtained from the intercept ($= 2\nu_0$).

(d) B_e and α_e are calculated from equation (4), and r_1, r_0 and r_e are calculated using equation (3).

(e) ω_e and $\omega_e x_e$ are calculated using the information that the first overtone of HCl^{35} has its band origin at 5668·0 cm^{-1}.

(f) From equation (2) the energies (cm^{-1}) of the rotational levels of the $v = 0$ and $v = 1$ vibrational states are calculated. An energy level diagram is constructed in which the rotational levels for the two vibrational states are drawn side by side. All the P and R transitions observed in the spectra are drawn and labelled.

[1] *Molecular Spectra and Molecular Structure*, Vol. I. *Spectra of Diatomic Molecules*, G. HERZBERG. D. van Nostrand Co., London, 1950.

[2] *Tables of Wavenumbers for the Calibration of Infra-Red Spectrometers*, I.U.P.A.C., Butterworths, London, 1961.

The Dissociation Energy of Iodine from its Absorption Spectrum

A. G. BRIGGS,

Loughborough University of Technology

Discussion

Iodine vapour absorbs visible radiation to become electronically excited. Absorption takes place largely from the vibrational level $v'' = 0$ in the ground electronic state to a series of vibrational levels in the upper electronic state. The resulting absorption band progression converges to a continuum corresponding to dissociation of the molecule (Fig. 1). In practice, the precise convergence limit cannot be observed

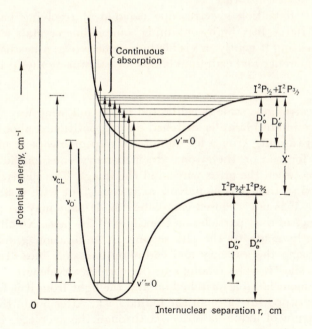

FIG. 1. Potential energy curves for the iodine molecule.

but has to be obtained by an extrapolation method. At higher temperatures especially, there may be an appreciable population of the $v'' = 1, 2, \ldots$ levels resulting in other progressions being superimposed on the $v'' = 0$ progression.

From Fig. 1 it can be seen that the convergence limit ν_{CL} is given by

$$\nu_{CL} = D_0'' + X = \nu_0 + D_0'$$

where X is the excitation energy of the $^2P_{\frac{1}{4}}$ iodine atom relative to the $^2P_{3/2}$ state and D_0 is the dissociation energy of an electronic state relative to the $v = 0$ level. A spectral study has shown that $X = 7599$ cm^{-1} and that $\nu_0 = 15,598$ cm^{-1}. If the convergence limit ν_{CL} is obtained, then values for D_0' and D_0'' may be deduced. A graphical extrapolation for ν_{CL} is preferable to the Birge–Sponer extrapolation which assumes that the vibrational term values, relative to the potential energy curve minimum, are given by

$$G(v) = \omega_e(v+\tfrac{1}{2}) - \omega_e x_e(v+\tfrac{1}{2})^2$$

For the excited electronic state of iodine, $\omega_e = 128$ cm^{-1} and $\omega_e x_e = 0 \cdot 834$ cm^{-1}. The above topics are fully covered by Herzberg[1] and Gaydon.[2]

Apparatus and Chemicals

Constant deviation spectroscope capable of resolving bands 30 Å apart at 5100 Å, litre bulb containing some iodine crystals with provision for heating it gently, a visible light source (e.g. a headlamp bulb) and lens, mercury and cadmium lamps for calibrating the spectroscope.

Method

The spectroscope scale is calibrated in the following way. The mercury (or cadmium) lamp is switched on and positioned about 2 inches in front of the slit. After a few minutes the spectroscope is adjusted so that it is focused on the strong green mercury emission line with no parallax between the cross wires and the image. The spectroscope slit is adjusted so that it is as narrow as is consistent with a strong even image, but it is never closed completely as damage may be caused by the presence of dust particles on the edges of the jaws. No alteration is subsequently made to the slit setting. A calibration graph is constructed using the mercury and cadmium emission lines (Table 1) to relate the spectroscope readings to the correct wavelength.

The filament lamp is switched on and the light from it is focused on the spectroscope slit through the iodine flask. After 5 min the absorption spectrum of iodine is observed through the eyepiece. It consists of a series of bands each with a sharp and diffuse edge. The diffuseness or shading is caused by unresolved rotational structure. If necessary, the temperature of the flask is gradually raised until the bands are clear. Too high a temperature will produce undesirable extra bands originating from the $v'' = 1$, or higher levels.

TABLE 1. MERCURY AND CADMIUM EMISSION LINES
(All the lines are fairly strong except those labelled (w))

Source	Wavelength Å	Colour region
Hg	4358·3	violet
Cd	4678·1 (w)	blue
	4799·9	
Hg	4916·0 (w)	blue–green
	4960·3 (w)	
Cd	5085·8	green
Hg	5460·7	
	5769·6	yellow
	5790·7	
Cd	6438·5	red

The wavelengths of the band heads are measured, starting with the one near 5430 Å (the $v'' = 0$ to $v' = 27$ transition) and the spectroscope is always adjusted in the same direction so as to avoid backlash errors. Each band head is measured at least twice and a mean value is calculated. If possible, bands up to a v' value of 50, or higher, are measured. The mean wavelength of each band head is converted into wavenumber units (cm^{-1}) *in vacuo* using five figure reciprocal tables, a calculating machine or wavelength–wavenumber conversion tables.[3] If conversion tables are not available the following corrections are applied to the observed wavelengths to allow for the refractive index of air.

TABLE 2. WAVELENGTH CORRECTION FACTORS

λ_{air}	Add	λ_{air}	Add
4000	1·13	5600	1·55
4400	1·23	6000	1·66
4800	1·34	6400	1·77
5200	1·44	6800	1·87

The first differences $\Delta\nu$ of the wavenumber values of the observed band heads are plotted against the value of v'. ν_{CL} is found by extrapolating the curve to $\Delta\nu = 0$ and adding to the wavenumber of any convenient band the area under the curve between this band and the intercept with the abscissa. D_0' and D_0'' (cm^{-1}) are calculated using the equations given in the discussion and are converted to electron-volts and cal mole^{-1}.

$$1 \text{ eV} = 8067 \cdot 5 \text{ cm}^{-1} = 23{,}053 \text{ cal mole}^{-1}.$$

[1] G. HERZBERG, *Molecular Spectra and Molecular Structure*. Vol. I. *Spectra of Diatomic Molecules*, van Nostrand, 1950.

[2] A. G. GAYDON, *Dissociation Energies and Spectra of Diatomic Molecules*, Chapman & Hall, 1953.

[3] C. D. COLEMAN *et al.*, *Table of Wavenumbers* (1960), **1**, National Bureau of Standards, Washington, D.C.

The Interpretation of Nuclear Magnetic Resonance Spectra

J. M. WILSON and A. G. BRIGGS,
Loughborough University of Technology

Discussion[1]

A spinning spherical particle of mass m and charge e behaves as an elementary magnet (Fig. 1).

FIG. 1. Model of spinning charged particle.

The ratio of its magnetic moment μ and its angular momentum L is given by

$$\mu/L = e/(2mc) = \gamma \qquad (1)$$

where c the velocity of light is the ratio between the electrostatic and electromagnetic units of measurement (3×10^{10} cm sec^{-1}). The ratio γ is called the magnetogyric ratio.

In the presence of a magnetic field the charged particle will not remain as shown in Fig. 1. It will experience a torque T whose direction is perpendicular to both μ and the magnetic field H_z and whose magnitude is given by

$$T = \mu H_z \sin \theta \qquad (2)$$

where θ is the angle shown in Fig. 2. The particle will precess about the applied magnetic field as shown in Fig. 2 (Larmor precession).

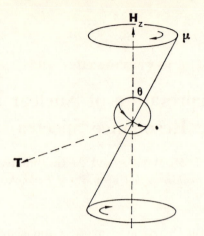

FIG. 2. Larmor precession.

The rate of change of $\mathbf{L}$ with respect to t is given by

$$d\mathbf{L}/dt = \boldsymbol{\mu} \times \mathbf{H}_z \quad \text{(vector product)} \tag{3}$$

and as

$$\boldsymbol{\mu} = \gamma\mathbf{L}$$

$$d\boldsymbol{\mu}/dt = \gamma(\boldsymbol{\mu} \times \mathbf{H}_z) = -\gamma(\mathbf{H}_z \times \boldsymbol{\mu}) \tag{4}$$

$$= \boldsymbol{\omega}_z \times \boldsymbol{\mu} \tag{5}$$

where $\boldsymbol{\omega}_z$ is called the Larmor precession frequency.

In the presence of mutually perpendicular external magnetic fields, $\mathbf{H}_z$ and $\mathbf{H}_1$, the particle will experience two torques (see Fig. 3). If the

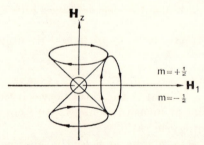

FIG. 3. Spinning particle in mutually perpendicular external fields.

second torque is such that its magnitude is equal to $\boldsymbol{\mu} \times \mathbf{H}_1$ (Larmor frequency $\boldsymbol{\omega}_1$) this torque will change as $\boldsymbol{\mu}$ changes with $\mathbf{H}_1$. This is true for all values of these two Larmor frequencies except when

$$\boldsymbol{\omega}_z = \boldsymbol{\omega}_1 \tag{6}$$

At this point $\boldsymbol{\mu}$ cannot simultaneously rotate about both axes and there is an exchange from one level to another as shown in Fig. 3. This exchange is referred to as resonance and in the case of nuclei, nuclear magnetic resonance.

With the exception of the hydrogen nucleus, nuclei consist of a number of protons (p) and of neutrons (n). The protons possess mass, charge and spin $\frac{1}{2}$. The neutrons differ in that they do not possess charge. The total mass is therefore $p+n$, the total charge p, and the spin a vector combination of $p+n$ spins each of magnitude $\frac{1}{2}$. Hence nuclei may be grouped as in Table 1. Of these three classes, only class A nuclei have no resonance spectra. Many of the common elements possess nuclei of this class but this is not disadvantageous. On the contrary it is advantageous, e.g. in a compound consisting of carbon, oxygen and hydrogen the proton signal is the only one obtained and is not complicated by additional fine structure caused by the presence of the other nuclei.

TABLE 1

Class	p	n	$p+n$	s	Examples
A	even	even	even	0	^{4}He, ^{12}C, ^{16}O
B	odd	odd	even	integral	^{2}H, ^{14}N
C			odd	half integral	^{1}H, ^{11}B, ^{13}C, ^{19}F

Note the spin of the nucleus is given the symbol I, the spin quantum number, and according to quantum theory the angular momentum is $\sqrt{\{I(I+1)\}}(h/2\pi)$ or $\sqrt{\{I(I+1)\}}$ a.m. units.

Relationship between Spin and Magnetic Field

The concluding paragraph of the previous section and equation (1) give the relationship

$$\mu = (e/2mc)\mathbf{I} = \frac{e\sqrt{\{I(I+1\}}}{2mc} \cdot \frac{h}{2\pi} \tag{7}$$

Strictly, this relationship is valid only if the nucleus is regarded as a point charge. For a real nucleus a dimensionless constant g, characteristic of each nucleus, has to be included, i.e.

$$\mu = g \cdot \frac{e\sqrt{\{I(I+1)\}}}{2mc} \cdot \frac{h}{2\pi} \tag{8}$$

which when expressed in nuclear magnetons β_N becomes

$$\mu = g\beta_N\sqrt{\{I(I+1)\}} \tag{9}$$

where

$$\beta_N = \frac{eh}{4m\pi c} = 5 \cdot 050 \times 10^{-24} \text{ erg gauss}^{-1}$$

When a nucleus with spin quantum number I is placed in a magnetic field H_z the magnetic quantum number M_I can take one of $(2I+1)$ values. There will therefore be $(2I+1)$ magnetic values with energy given by

$$\Delta E = -g\beta_N H_z M \tag{10}$$

The magnetic dipole selection rule for transitions between these levels requires that ΔM must always be unity, hence even for nuclei where I is greater than $\frac{1}{2}$ there is only one resonance absorption frequency (see Fig. 4).

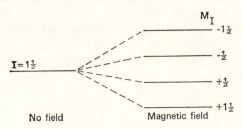

FIG. 4. Splitting of nuclear energy levels in a magnetic field.

The energy difference between the levels ΔE is given by

$$\Delta E = E^{(2)} - E^{(1)}$$

$$= \mu_z{}^{(1)} - \mu_z{}^{(2)}$$

$$= g\beta_N I_z{}^{(1)} H_z - g\beta_N I_z^2 H_z$$

$$= g\beta_N H_z \text{ ergs} \tag{11}$$

where H_z is expressed in gauss. But

$$\Delta E = h\nu \tag{12}$$

where ν is the frequency, i.e.

$$\nu = \frac{g\beta_N H_z}{h} \text{ c/s} \tag{13}$$

which for a field of 14,000 gauss and a constant $g = 5 \cdot 585$ (for a proton) gives

$$\nu = \frac{5 \cdot 585 \times 5 \cdot 05 \times 10^{-24} \times 1 \cdot 4 \times 10^4}{6 \cdot 63 \times 10^{-27}}$$

$$\approx 60 \times 10^6 \text{ c/s}$$

Hence for a magnetic field of 14,000 gauss a frequency of approximately 60 megacycles/sec is required for the nuclear magnetic resonance spectra of ^{1}H. Other approximate frequency values for nuclei of particular interest are given in Table 2.

TABLE 2

Isotope	Frequency (Mc/s)
^{1}H	60·0
^{2}H	8·7
^{11}B	19·25
^{13}C	15
^{14}N	4·3
^{19}F	56

Chemical Shift

Assemblies of nuclei of these isotopes will therefore absorb energies at the frequencies given in the previous section and an obvious application would be the comparison of the magnetic moments of the nuclei by comparing their resonance frequencies in the same applied field. A particular isotopic species in the applied field gives, as expected, sharp resonance lines, but the value of the relative positions of these lines depends upon the chemical environment of the nucleus. This discovery is of immense significance to the chemist, for it enables him to identify nuclei in different chemical environments by the displacement or shift of the lines, and the phenomenon is given the name "the chemical shift".

The chemical shift is due to the shielding of the nucleus by its surrounding electrons, i.e. there is a screening effect from the magnetic field. In an applied magnetic field H_z the electrons give rise to circulating electric currents which produce an induced magnetic field H' at the nucleus and opposite to the applied field. Hence the magnitude of the effective field H is given by

$$H = H_z - H' \tag{14}$$

$$= (1 - \sigma)H_z \tag{15}$$

where σ is a dimensionless number known as the screening constant. For two nuclei of the same isotopic species but in different environments i and j, the chemical shift difference δ_{ij} is given by

$$\delta_{ij} = \sigma_i - \sigma_j \tag{16}$$

The magnitude of this difference for proton resonance is of the order of magnitude of 10 parts per million (ppm).

The absolute measurement of this chemical shift is possible but for comparisons between different instruments its measurements relative to a standard are made. Since few protons are shielded more than those in tetramethyl silane (TMS) their signal is a suitable reference one. The relative chemical shift δ_{sr} will be given by

$$\delta_{sr} = \frac{H_s - H_r}{H_r} \qquad (17)$$

where H_s and H_r are the fields corresponding to the resonance of a particular proton in the compound being investigated and the TMS protons. Since field strengths are proportional to frequencies (ν) we have

$$\delta_{sr} = \frac{\nu_s - \nu_r}{\nu_r} \qquad (18)$$

But as previously noted for protons, the magnitude of the chemical shift difference is of the order of 10 ppm, hence if we multiply the ratio by 10^6 we will obtain a more convenient number expressed in ppm

$$\delta \, (\text{ppm}) = \frac{\nu_s - \nu_r}{\nu_r} 10^6 \qquad (19)$$

This number will be a negative one and to avoid the inconvenience of working with negative numbers Tiers introduced the quantity τ (tau)

$$\tau \, (\text{ppm}) = 10 + \delta \qquad (20)$$

On the tau scale TMS is assigned a value of 10·00. For protons in the usual types of electron environments (with the important exception of acidic protons) tau values are therefore positive and the usual tau range of ten units extends over 600 c/s at a field strength of 14,092 gauss (60 Mc/s). On a 100 divisions chart a resonance signal separated by 10 divisions from TMS will therefore be 60 c/s from the latter and the tau value will be

$$\tau = 10 - \frac{60 \times 10^6}{60 \times 10^6}$$

$$\tau = 9$$

Nuclear magnetic resonance spectral positions for protons in some organic structures are given in Chart 1.

The τ value for the proton in a carboxylic acid cannot be accommodated on the 0 to +10 tau scale. If present its signal will appear in an inset on the chart where the scale for the inset has been displaced 200 c/s in the direction of lower field. The usual scale extends over a

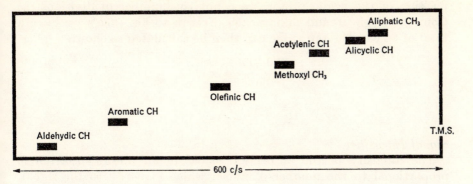

CHART 1. CHARACTERISTIC nmr SPECTRAL POSITIONS FOR PROTONS IN SOME
ORGANIC STRUCTURES

range of 0 to 10 tau units or 600 c/s at 60 Mc/s. Within this range of
values the calibrations on the preprinted chart are valid. The chart
scale is divided into 100 divisions and the usual scale is as shown in
Fig. 5(a).

The magnetic field can be applied, e.g. to accommodate scanning at
800 c/s from the TMS resonance position. The preprinted calibrations
do not hold under these circumstances, and when the usual chart
paper is used in the apparatus the additional range of 200 c/s can only
be shown as an inset. This is most conveniently shown by placing the
start of the offset scale (0′) above the origin 0 on the chart as shown
in Fig. 5(b). Note carefully that the negative sign merely indicates
that the origin is 800 c/s from the TMS signal. In order to locate the

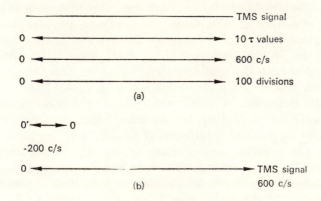

FIG. 5. (a) Usual chart scale; (b) chart scale with offset scale.

position of the acidic proton we count the number of divisions (n) from the origin 0′ and taking each division to be $+6$ c/s we have a total of $6n$ c/s. The tau value can then be calculated as shown

$$\frac{800 - 6n}{60} = x$$

$$\tau = 10 - x$$

Signal Intensities

The intensity of the absorption signal in an n.m.r. spectrum is proportional to the number of nuclei which give rise to the signal. Hence the determination of relative peak intensities provides information on the number of different proton (or other) environments in a molecule and hence on molecular structure. Relative intensities of the bands in a given spectrum can only be obtained by a comparison of peak areas. Peak heights can be used but are not reliable unless the bands are of the same shape and are symmetrical. Some n.m.r. spectrometers have circuitry which enables peak areas to be evaluated electronically, the peak area being proportional to the vertical displacement between two lines drawn by the recorder pen on the chart at a point corresponding to the resonance frequency.

Spin–Spin Coupling

For acetaldehyde the low resolution spectrum consists of two peaks (areas 3 : 1) as expected. In a high resolution study the peaks are seen to possess fine structure, the smaller peak consisting of four minor peaks, and the larger one of two minor peaks. It is information of this sort which can be very valuable in identifying structural units of the molecule. The origin of the fine structure lies in spin–spin coupling, an effect which is usually found to apply to protons attached to adjacent atoms.

Briefly, the proton resonance of the methyl group is split into two by interaction with the single aldehydic proton. This proton can have orientations $+\frac{1}{2}$ or $-\frac{1}{2}$ and interaction with each of these gives rise to a slightly different resonance; the methyl proton resonance peak is therefore split into a doublet. On the other hand, the aldehydic proton "sees" three equivalent neighbouring protons (on the methyl group) for which the possible combinations of orientation are as shown in Fig. 6 where the spins are indicated by α and β respectively.

The aldehydic resonance line is therefore a closely spaced quartet, the intensities of the sub-peaks being in the ratio 1 : 3 : 3 : 1. (A similar effect is found for the action of the F nucleus on the P nucleus

in PF_3.) It should be noted that the ratio of the areas of the peaks is still
1 : 3 whether the high-resolution or the low-resolution spectra are
considered.

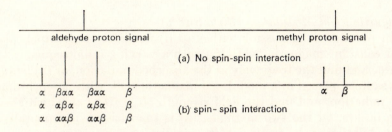

Fig. 6. Representation and interpretation of the n.m.r. spectrum of
acetaldehyde.

This magnetic interaction between neighbouring nuclei is known as
spin–spin coupling. It is transmitted by the valence electrons associated with the nuclei and is represented by the symbol J, the magnitude
of which is dependent on the degree of delocalization of the electrons
between the nuclei. J in contrast to δ is independent of the magnitude
of H_z. Note when J is small compared with the chemical shift difference,
and when each nucleus in one group is equally coupled with each
nucleus in the other group (as in the case of acetaldehyde), the multipli-
city for one group of protons is given by the $(n+1)$ rule where n is the
number of protons in the adjacent group. The intensities of the components are given by the coefficients of the expansion of $(1+x)^n$.

Apparatus

A nuclear magnetic resonance spectrometer is an extremely complex
instrument but consists of four basic sections, viz. the source of magnetic field, a means of modulating this field over a tiny but precise
range, a radiofrequency oscillator and a detector.

The Magnetic Source

In n.m.r. spectrometry, a magnet gap of about $1\frac{1}{2}$ inches is customary,
and to ensure homogeneity of the field a poleface diameter to gap ratio
of 5 : 1 is necessary. (A homogeneous field is defined as one in which
the strength and direction of the field do not vary from point to point.)
In the Perkin–Elmer spectrometer (model R10) a permanent magnet of
14,000 gauss is used. This has the advantage that it needs no major
power supply and no cooling system. As the temperature coefficient of

L*

the field of a permanent magnet is about -2 in $10^4/°C$ accurate temperature control is essential and in practice the magnet is kept in a special insulated enclosure which is thermostatted at a temperature about 15°C above ambient.

Magnetic Field Modulation (Field Sweep)

In practice, it is possible either to (i) irradiate the sample, placed in a constant magnetic field (H_z) with a varying radiofrequency (RF) source and find the frequency of the absorbed radiation, or (ii) irradiate the sample, placed in a modulated magnetic field, with monochromatic RF radiation, and find the field strength ($H_z \pm \Delta H_z$) at which absorption occurs. Of the two methods, (ii) is the easier experimentally.

Modulation of the static field H_z may be brought about in two ways. Firstly, a suitably varying current may be passed through auxiliary coils wound around the pole pieces of the magnet, causing the effective value of H_z to vary. Secondly, varying current is sent through a pair of identical Helmholtz coils suspended in the gap between the pole pieces and separated by a distance equal to a coil radius. In either case, field homogeneity can be maintained by careful design, and the rate of change of the sweep field can be made to be constant with time (i.e. linear sweeps). The Helmholtz coil system is capable of generating higher sweep frequencies because of the much lower circuit inductance.

The Radiofrequency Oscillator

With the magnetic fields commonly in use (about 14,000 gauss) it is found that radiation in the radiofrequency region is absorbed by suitable samples. Some typical resonance frequencies are given in Table 2 and for a magnetic field of about 14,000 gauss, a frequency of 60 Mc/s is required for protons.

It is not feasible to generate stable frequencies of about 60 Mc/s directly, and the usual procedure is to drive a frequency multiplier from a highly stable, temperature controlled 5 Mc/s crystal oscillator. The final RF signal, the power of which can be varied to suit the sample, is applied to a sample coil situated in the pole gap of the magnet and wound with its axis perpendicular to the direction of the magnetic field. The resulting linearly oscillating field can be shown to be resolvable into two components (of magnitude H_1) which are rotating in phase but in opposite directions. One of the components will be rotating in the same sense as the precessing nuclear magnets of the sample, with which interaction will occur at the resonance frequency. The second component need not be considered further since it has no effect on the nucleus.

The Detector System

A means has to be devised for detecting the absorption of energy from the RF source. In the bridge method the RF voltage arriving at the detector (diode) amplifier is reduced by balancing out most of the output from the source by means of a radiofrequency bridge. A typical system is shown in Fig. 7.

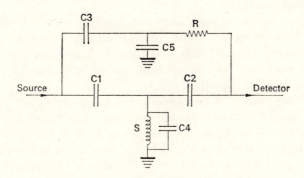

Fig. 7. Radiofrequency detector bridge.

The RF voltage is changed in phase 180° at each of the capacitors C_1 and C_2, but as the capacitance of C_4 is changed slightly, the phase of the voltage across S is changed. The phase of the voltage passing through C_2 is therefore about 0 degrees with respect to the source but variable slightly on either side by means of C_4. The RF passing along the other arm of the bridge through R and C_3 is changed in phase by only 180 degrees, but can be changed in amplitude by varying C_5 in the voltage divider formed with C_3. The currents flowing through the two arms of the bridge can therefore be changed in amplitude (with C_5) and phase (C_4) so that they just cancel one another. Any change of voltage, due to nuclear resonance (sample absorption), produced across S, and passed through C_2 to the amplifier, therefore appears as a significant part of the input to the amplifier, and so a much better signal-to-noise ratio is obtained.

The Sample

Even under the best conditions attainable in practice, there may still result a small degree of inhomogeneity, partly arising from a non-perfect magnet and partly from the finite size of the sample. Improved resolution can be obtained by rotating the sample ("spinning") at about 1500–2000 rev/min. This is much faster than the sweep rate of the field H_1 and is achieved by fitting a miniature turbine wheel on the bottom of the sample tube. Compressed air (10 to 20 lb/in²) is used to rotate the turbine wheel/sample tube assembly. The effect of spinning the sample is to subject the sample molecules (at least those in a given horizontal plane) to an average magnetic field. Care should be taken to remove solid particles from a liquid sample, since a drift of the particles through the volume of the liquid can adversely affect the field homogeneity.

Most high-resolution spectra are obtained from pure liquid samples, or, in the case of solids, from solutions. Solid samples give broad absorption bands (loss of fine structure) and gaseous samples, unless under high pressure, do not contain enough material to produce an adequate absorption signal.

Samples are placed in precision glass tubes, typically some 10 to 15 cm long and 2 mm in internal diameter. The tubes are filled to a depth of about 3 to 4 cm, corresponding to a volume of sample of approximately 0·5 cm³. If a slight loss of resolution is acceptable, then a small plug (nylon) may be placed in the sample tube thereby enabling a smaller volume of liquid to be used.

When solutions are being used the possibility of solvent–solute interactions should be remembered; these interactions may cause a chemical shift to be proportional to the solute concentrations. Concentrated solutions can give rise to line broadening and loss of resolution. This phenomenon also occurs with viscous media and is caused by incomplete averaging of dipole–dipole interactions. The remedy is to dilute the sample. For a molecule containing a single proton, the limit of detection will be reached at a molarity of about 0·02. Working concentrations should be much higher than this, of course.

Solvents

The requirements of a solvent in n.m.r. spectroscopy are that it should enable a fairly concentrated, low viscosity solution to be made and should not interact with the solute; there should be no solvent resonances in the region of spectral interest. The most widely used solvents are carbon tetrachloride, carbon disulphide, chloroform, deuterochloroform, methylene chloride, dioxan, cyclohexane, and, to a lesser extent, benzene, pyridine, acetone, dimethylformamide and acetonitrile. Carbon tetrachloride should be used whenever possible as it is magnetically isotropic.

Temperature Control

Since the sample (liquid, in a narrow-bore glass tube, either sealed or corked) will normally be at room temperature, then it must be pre-heated to the approximate temperature of the magnet housing before being inserted into the magnet gap, If this is not done, or if too many samples are inserted in rapid succession, the temperature of the magnet may fluctuate sufficiently to have an adverse effect on the resolution.

Presentation of Spectra

It has been mentioned previously that one can look for an absorption peak by (i) irradiating the sample with a varying radiofrequency source, maintaining the magnetic field (H_z) at a constant value, (ii) irradiating the sample with a constant radiofrequency source, the magnetic field (H_z) being varied over a small range. Most instruments use method (ii), although the abscissae of the recorder charts are calibrated as though method (i) were being used. Spectra are automatically drawn on a recorder chart and there is usually provision for an oscilloscope presentation as well. Ordinates correspond to absorption in arbitrary units.

The ideal reference substance would need the following properties. It should be chemically inert, give a single, narrow, intense absorption signal, be miscible with a large range of organic compounds (and solvents) and should preferably be fairly volatile in order to make the recovery of small, valuable samples feasible. Currently, the most suitable reference compound for proton resonances is tetramethylsilane, $Si(CH_3)_4$. There are twelve symmetrically arranged identical protons in TMS which give rise to a peak very suitable for reference purposes. In addition, the molecule of TMS is magnetically isotropic. TMS has a further advantage as a reference compound in that its absorption line lies at a higher frequency than those of most other "organic" protons, i.e. most proton resonances lie on the same side of the reference line and there is no problem of $+$ and $-$ displacements. Usually, it is sufficient to add about 0·2 per cent of TMS (b.p. 27°C) to the liquid sample, since intermolecular interactions are unlikely to be sufficiently large to affect the standard resonance frequency.

For the purpose of this investigation a number of RECORDED SPECTRA are AVAILABLE. Students must not operate the spectrometer.

Method

SEC. I. *Number of resonance lines*

The recorded spectra should be viewed as a whole. Sufficient experience may enable a *provisional* assignment of peaks to be made by inspection, but a logical analysis should follow.

There is no difficulty when the recorded peaks are fairly widely spaced singlets. Such peaks correspond to groups of protons in particular environments, e.g. CH_3, —COOH, etc., and not undergoing inter-actions with nearby protons.

Peaks which occur as fairly closely spaced doublets usually arise when the proton(s) giving rise to the peak are adjacent to a single different proton. Discretion must be exercised in deciding whether a given doublet is a true doublet (arising from a single type of proton) or is simply two nearby singlet peaks (arising from two different types of non-interacting protons). In certain cases of doubt, each possibility must be considered in turn. True doublets, to a first approximation, often have more or less the same intensity.

Similar arguments apply to triplet and quartet peaks. For example, in the high resolution spectrum of ethanol, a quartet and a triplet occur. These correspond to the methylene and methyl groups respectively. The splitting can be readily understood provided it is remembered that the field experienced by the protons of one group is influenced by the spins of the protons in the *neighbouring group*. Application of the $(n + 1)$ rule indicates quartet and triplet multiplets. Again it should be remembered that multiplet structures generally have a characteristic appearance, i.e. the peaks are equally spaced and have the ideal intensity ratios, $1 : 2 : 1$ for a triplet and $1 : 3 : 3 : 1$ for a quartet.

Having made a preliminary study the spectra should be viewed in the light of the recorded intensities and chemical shifts. This may or may not confirm the initial conclusions and a completely false analysis may have been made due to

    (a) inadequate resolution of peaks,
    (b) the masking of small peaks by the solvent,
    (c) the "overlooking" of peaks by the observer,
    (d) the presence of impurity.

SEC. II. *Intensities*

The intensities are related to the areas under the peaks of particular resonance signals. These peak areas are evaluated electronically by the spectrometer, each being proportional to the vertical displacement between the lines drawn by the recorder pen on the chart. The divisions covered by the trace for all the displacements are counted and the number obtained is divided by the total number of hydrogen atoms present in the molecule. In this way the average number of divisions for each hydrogen atom can be calculated. Hence the number of hydrogen atoms occurring in a particular environment can be found. Note the above statement is not valid for the TMS peak.

SEC. III. *Chemical shift*

If the 0 to 10 scale on the base of the chart (usually labelled ppm) corresponds to τ (tau) values directly, no conversion is necessary. This scale is used to determine the tau value of each assigned peak. In the case of multiplets, the value is read from the "centre of gravity" (this is also the mid-point of the multiplet for simple spectra). Tau values need only be read to the nearest 0·1 unit, in the first instance.

From the correlation chart and tables provided, provisional assignments of possible structures giving rise to each resonance peak should be made. The assignments should be consistent with the molecular formula of the compound.

The individual groups present should then be combined to give a structural formula and it is essential that this formula is such that it would be expected to yield the observed spectra in every detail.

Examples

The first four charts record the n.m.r. spectra of
(1) ethanol in carbon tetrachloride (20 per cent v/v),
(2) ethanol in chloroform (20 per cent v/v),
(3) ethanol in chloroform (50 per cent v/v),
(4) pure ethanol in carbon tetrachloride (20 per cent v/v).

The student should study Chart (1) carefully and note the number of resonance lines. The number and grouping of these lines should be considered as indicated in Sec. I. The splitting of the signal from the methyl group by spin–spin interaction with the methylene protons, and the corresponding splitting of the signal from the methylene group by the methyl protons are as in Fig. 8(a) and (b), respectively. The value of the coupling constant J should be carefully measured with a pair of dividers and a ruler marked in millimetres. The result obtained should be converted to cycles per second.

The intensities of the signals should be estimated as directed in Sec. II. Finally the value of the chemical shifts should be measured directly from the tau scale and deduced from the position of the resonance lines and the value of J.

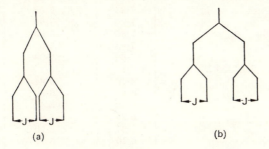

(a)   (b)

FIG. 8. Splitting of the signal for (a) methyl protons and
(b) methylene protons in ethanol.

Results and conclusions should be summarized in tabular form.

Tau value	H	Multiplicity	Assignment
	3	Triplet	CH_3 (a)
	2	Quartet	CH_2 (b)
	1	Singlet	OH (c)

Molecular formula

$$CH_3 \cdot CH_2 \cdot OH$$

(a)   (b)   (c)

Value of coupling constant =   c/s

Tau values deduced from Fig. 8 =

Charts (2) and (3) should be studied and in particular the influence of the solvent on the number of resonance lines and on their positions should be noted and discussed.

Chart (4) should be studied and the increase in the number of resonance lines should be noted and explained. (Spin–spin interaction between groups (a) and (c) should be neglected.) The value of the coupling constants between methyl and methylene groups (J_1) and between methylene and hydroxyl groups (J_2) should be deduced. The position of the chemical shifts as measured directly and as deduced from spin–spin interactions should be recorded.

Twelve other charts are provided and these should be discussed in a similar way, omitting the determination of the value of coupling constants.

TABLE 3. VALUES (± 0.5 tau units) FOR THE PROTONS OF CH_3, CH_2 AND CH GROUPS
(Attached to various groups X and to saturated hydrocarbon residues, R, R', etc.)

X	$CH_3.X$	$R'CH_2.X$	$R'R''CH.X$
—R	9·10	8·75	8·50
—CO.R	8·00	7·90	
—CN	8·00	7·52	
—$CO.NH_2$, —$CO.NR_2$	7·98	7·95	
—CO.OH	7·93	7·66	7·43
—CO.R	7·90	7·60	7·52
—NH_2, —NR_2	7·85	7·50	7·13
—CHO	7·83	7·8	7·6
—Ph	7·66	7·38	7·13
—NH.CO.R, —$NR'.CO.R''$	7·1	6·7	6·5
—OR	6·70	6·64	6·20
—NR^+_3	6·67	6·60	
—OH	6·62	6·44	6·15
—O.CO.R	6·35	5·85	4·99
—O.Ph	6·27	6·10	6·0
—O.CO.Ph	6·10	5·77	4·88
—NO_2	5·67	5·60	5·40

Reproduced by permission of the Royal Institute of Chemistry and Academic Press.

Tau Values for OH *and* NH *Groups*

R.OH

The value of τ varies considerably from 4·0 to 9·0. In general it is lower for enols, and hydrogen bonded enols can be as low as -1 to -6. Theoretically a hydroxyl group should give a multiplet due to coupling with adjacent protons but experimentally it is usually observed as a singlet. In the presence of a trace of acid this singlet is sharpened, e.g. for R.OH and water it is approximately 5·3. This value is due to the catalysed exchange between the alcohol and the solvent. The position of the signal varies with concentration, temperature and the nature of the solvent except for non-polar solvents where it is essentially independent of the concentration. The great variation of the position of the signal for the hydroxyl proton often makes it difficult to identify. Even if the observer is not interested in the identification of this group its presence may cause confusion and this line may be assigned to some other structural feature. This difficulty may be overcome by taking advantage of the temperature and concentration variation of the hydroxyl group. It is, however, preferable to replace the hydrogen atom of the group. This can be done by shaking the compound with deuterium oxide.

R.NH$_2$

Again exchange with solvent is possible and the position of the singlet depends upon the concentration and nature of the solvent. The singlet is sharpened by the addition of a trace of acid due to the catalysed exchange. This is not true in the presence of concentrated acid where in the case of R.NH$_2$ protonation is complete, giving R.NH$_3^+$. This suppresses the exchange and the signal disappears.

Solvents

The approximate line positions for some solvents are given in Table 4.

TABLE 4

Solvent	τ value
Chloroform	2·73
Benzene	2·73
Acetone	7·83
Dioxan	6·3
Water	5·3

The intense solvent signal is always accompanied by spinning side bands. For many organic substances carbon tetrachloride is the best solvent. For water or deuterium oxide solutions dioxane, acetonitrile ($\tau = 8\cdot 0$), and trimethyl silane ethyl sodium sulphonate ($\tau = 10\cdot 0$) can be used.

A number of excellent textbooks are available:

Nuclear Magnetic Resonance Spectra, L. M. JACKMAN, Pergamon Press, 1962.

High Resolution Nuclear Magnetic Resonance Spectroscopy, POPLE, SCHNEIDER and BERNSTEIN, McGraw-Hill, 1959.

High Resolution Nuclear Magnetic Resonance, EMSLEY, FEENEY, and SUTCLIFFE, Pergamon Press, 1966.

Fundamentals of Molecular Spectroscopy, BANWELL, McGraw-Hill, 1966.

Nuclear Magnetic Resonance, C. D. and O. JANDETSKY.

Comprehensive Biochemistry, edited by FLORKIN and STOTZ, Vol. 3, Elsevier, 1962.

Explosion Limits of the Hydrogen–Oxygen Reaction

A. N. STRACHAN,

Loughborough University of Technology

Discussion

The combination of hydrogen and oxygen is a chain reaction in which the chain centres are hydrogen and oxygen atoms and hydroxyl radicals. The reaction is almost certainly started by the formation of hydrogen atoms but the details of this initiation process are still in doubt.

Initiation $\qquad ? \to H \qquad$ Rate $= R_1$

The propagation steps are thought to be

Propagation $\qquad H + O_2 \to OH + O \qquad$ Rate $= R_2 = k_2[H][O_2]$

$\qquad\qquad\qquad O + H_2 \to OH + H \qquad$ Rate $= R_3 = k_3[O][H_2]$

$\qquad\qquad\qquad OH + H_2 \to H_2O + H \qquad$ Rate $= R_4 = k_4[OH][H_2]$

of which the first two are also branching reactions. At pressures near atmospheric the main termination step is thought to be

Termination $\quad H + O_2 + M \to HO_2 + M \qquad$ Rate $= R_5 = k_5[H][O_2][M]$

M is any other molecule present and plays the role of removing the excess energy from the HO_2 radical, thereby rendering it inactive.

For the reaction to proceed at a steady non-explosive rate, the chain centre concentrations must remain constant.

$$\text{If [O] is constant} \qquad R_2 = R_3 \qquad (1)$$

$$\text{If [OH] is constant} \qquad R_2 + R_3 = R_4 \qquad (2)$$

$$\text{If [H] is constant} \qquad R_1 + R_3 + R_4 = R_2 + R_5 \qquad (3)$$

Combining equations (1), (2) and (3)

$$R_1 + 2R_2 = R_5$$

or Initiation rate + Branching rate = Termination rate.
Hence

$$R_1 = R_5 - 2R_2$$
$$= (k_5[\text{M}] - 2k_2)[\text{H}][\text{O}_2]$$

The overall rate of the reaction is equal to the rate of water formation, so that

$$\text{Overall rate} = R_4 = 2R_2 = 2k_2[\text{H}][\text{O}_2] = \frac{2k_2 R_1}{k_5[\text{M}] - 2k_2} \tag{4}$$

It will be seen that a steady non-explosive rate is only possible if the termination rate is greater than the branching rate, i.e. if $k_5[\text{M}]$ is greater than $2k_2$. Reaction (2) is endothermic and as a result has a large activation energy and k_2 therefore increases rapidly with temperature. At temperatures above 600°C and at normal pressures $2k_2$ is larger than $k_5[\text{M}]$, a steady rate is never realized and the reaction is always explosive. At temperatures below 600°C and at atmospheric pressure, $k_5[\text{M}]$ is greater than $2k_2$ and the reaction proceeds at a slow steady rate given by equation (4). If the overall pressure is lowered, however, thereby reducing [M], a critical pressure will be reached at which $k_5[\text{M}] = 2k_2$. At this pressure the overall rate becomes infinite and an explosion occurs. This critical pressure is known as the second explosion limit. (The first explosion limit occurs at very much lower pressures when another termination step—the diffusion of the chain centres to the walls of the reaction vessel—results in the reaction becoming non-explosive once more.)

At the second explosion limit

$$k_5[\text{M}] = 2k_2$$

or

$$[\text{M}] = \frac{2k_2}{k_5} = \frac{2A_2}{A_5} \exp\left[-\frac{(E_2 - E_5)}{RT}\right]$$

Hence, the pressure P at which explosion occurs, being proportional to [M] should increase exponentially with temperature and a plot of $\log P$ against $1/T$ should be linear with a slope of $-(E_2 - E_5)/2 \cdot 303R$. Since E_5 is almost certainly zero, this plot should enable E_2, the activation energy of the chain branching reaction, to be determined.

If the reaction mixture consists of air and hydrogen, M can be either of these and the expanded form of $k_5[\text{M}]$ is $(k_{5A}[\text{Air}] + k_{5H}[\text{H}_2])$. At a given temperature explosion occurs when

$$k_{5A}[\text{Air}] + k_{5H}[\text{H}_2] = 2k_2 = \text{constant}$$

By determining the explosion pressure for different mixtures of air and hydrogen and by plotting P_{H_2} against P_{Air} (these being the partial pressures at the explosion limit), it should be possible to determine the ratio k_{5H}/k_{5A} and hence the relative efficiencies of hydrogen and air in removing energy from and stabilizing the HO_2 radical.

Apparatus and Chemicals

A diagram of the apparatus is shown in Fig. 1.

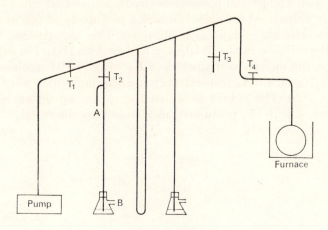

FIG. 1. The apparatus.

Method

The furnace temperature is adjusted until it is at a steady temperature between 450°C and 550°C. (It must not be allowed to rise above 570°C or the pyrex reaction flask will melt.) The pump is turned on and the whole system is evacuated for 15 min with T_2 and T_3 closed. After a hydrogen cylinder has been connected at A and the tubing AB flushed out, T_1 is closed and hydrogen is admitted to the system to a pressure of 200 torr by opening T_2 carefully. T_4 is then closed and the hydrogen which is not in the reaction flask is pumped away by opening T_1.

After closing T_1 again, air is admitted by opening T_3. T_4 is opened for a few seconds thus allowing air in on top of the hydrogen. The pressure of air in the reaction flask will then be atmospheric pressure less the hydrogen pressure.

After allowing 30 sec for mixing, T_3 is closed, T_4 is opened and the pressure gradually lowered by opening and adjusting T_1. The pressure in the apparatus when the explosion occurs (detected by a kick on the manometer) is recorded and a measurement of the temperature of the reaction flask is made at the same time. The procedure is repeated at several other temperatures in the range 450°–550°C using the same mixture of hydrogen and air each time. Plots are made of P against T and of $\log P$ against $1/T$ and an estimate of E_2 obtained.

The temperature is then held steady at some point between 500°–550°C and a number of runs carried out in which the hydrogen to air ratio is varied. Hydrogen pressures in the range 100–500 torr may be used. The partial pressures of hydrogen and air when explosion occurs are calculated for each run and a plot made of one against the other. An estimate of k_{5H}/k_{5A} is obtained. This is compared with the ratio of the collision rates of hydrogen and air with HO_2. For calculating the latter, the molecular diameters of H_2, air and HO_2 are assumed to be 2·7, 3·7 and 4·0 Å, respectively, and the molecular weight of air is taken as 29. The extent to which the results are in agreement or disagreement with the postulated mechanism are discussed.

Mechanism of the Reaction between Hydrogen Iodide and Hydrogen Peroxide

A. N. STRACHAN,

Loughborough University of Technology

Discussion

The overall reaction

$$H_2O_2 + 2HI \rightarrow 2H_2O + I_2$$

results from a sequence of steps. The slow rate determining step is thought to be the formation of HIO. A more rapid step follows to complete the reaction.

$$HIO + HI \overset{rapid}{\rightarrow} H_2O + I_2$$

$$(or \quad HIO + H^+ + I^- \rightarrow H_2O + I_2)$$

The most likely ways in which the HIO is formed are:

(a)
$$H_2O_2 + I^- \overset{k_1}{\rightarrow} OH^- + HIO$$

$$Rate = k_1[H_2O_2][I^-]$$

(b)
$$H_2O_2 + H^+ \underset{rapid}{\overset{k_2}{\rightarrow}} H_3O_2^+$$
$$\underset{k_3}{\overset{\leftarrow}{}}$$

$$k_2[H_2O_2][H^+] = k_3[H_3O_2^+]$$

$$H_3O_2^+ + I^- \underset{slow}{\overset{k_4}{\rightarrow}} H_2O + HIO$$

$$Rate = k_4[H_3O_2^+][I^-]$$

$$= \frac{k_4k_2}{k_3}[H_2O_2][H^+][I^-]$$

If the HIO is formed by both (a) and (b) then

Rate of overall reaction = Rate of HIO formation

329

and hence we would have,

$$\text{Rate of overall reaction} = k_1[H_2O_2][I^-] + \frac{k_4 k_2}{k_3}[H_2O_2][H^+][I^-]$$

$$= \left(k_1 + \frac{k_4 k_2}{k_3}[H^+]\right)[I^-][H_2O_2]$$

Apparatus and Chemicals

Thermostat, 20 vol. solution of hydrogen peroxide, 5 N sulphuric acid, 0·1 N sodium thiosulphate, 0·1 N potassium permanganate, starch solution, 250 ml graduated flask, conical flask, stop-clock.

Method

A dilute solution of hydrogen peroxide is prepared by taking 10 ml of the 20 vol. solution and making it up to 250 ml with distilled water. The flask containing this solution is placed in a thermostat at 25°C.

Five grams of potassium iodide are dissolved in 250 ml of distilled water and a 25 ml portion of this solution is pipetted into 100 ml of distilled water in a conical flask. After 10 ml of 5 N sulphuric acid and 5 ml of starch solution have been added this flask is also placed in the thermostat.

A burette is filled with 0·1 N sodium thiosulphate solution and clamped above the thermostat so that its contents can be discharged into the conical flask. Exactly 1 ml of sodium thiosulphate solution is then added from the burette.

When the contents of the conical flask have attained thermal equilibrium, the reaction is started by adding from a pipette 10 ml of the diluted hydrogen peroxide solution. The stop-clock is started as the peroxide is added and immediately the addition is complete the contents of the flask are swirled to mix them well.

When the solution turns blue, the time is noted (without stopping the clock) and a further 1 ml of thiosulphate solution is added from the burette. When the solution goes blue once more the time is again recorded and another 1 ml of thiosulphate solution is added. This procedure is repeated until 20–30 min have elapsed from the start of the reaction.

The dilute solution of hydrogen peroxide is standardized by adding 10 ml of 5 N sulphuric acid to a 10 ml aliquot of the peroxide and then titrating with 0·1 N potassium permanganate. If V ml of permanganate are required, an equal volume V ml of permanganate is placed in a flask, to which is added 10 ml of 5 N sulphuric acid and about 3 g of potassium iodide dissolved in 10 ml of distilled water. The liberated iodine is titrated with the 0·1 N sodium thiosulphate solution.

If 10 ml of peroxide is equivalent to V ml of permanganate which in turn is equivalent to A ml of thiosulphate, then when the reaction mixture first turns blue, the hydrogen peroxide concentration is equivalent to $(A-1)$ ml of thiosulphate. At the time when the solution turns blue for the second time, the peroxide concentration is equivalent to $(A-2)$ ml of thiosulphate, etc. A plot of $\log(A-n)$ against time should be linear with a slope of $-k/2\cdot303$ where k is the apparent first-order rate constant and according to the postulated mechanism should be given by

$$k = \left(k_1 + \frac{k_4 k_2}{k_3}[\mathrm{H^+}]\right)[\mathrm{I^-}]$$

Further runs, with the conditions suitably modified, are now carried out in order to investigate one or more of the following:

  (i) the dependence of k on $[\mathrm{I^-}]$
  (ii) the dependence of k on $[\mathrm{H^+}]$
  (iii) the dependence of k_1 and $k_4 k_2/k_3$ on temperature
  (iv) the dependence of $k_4 k_2/k_3$ on ionic strength.

In relation to (iv), of reactions (1) to (4), only reaction (4) is a reaction between two ions and is therefore the only one which should be influenced by the ionic strength I, of the solution. One would expect that

$$\log k_4 = \log k_4^0 + 2A\, z_1 z_2 \sqrt{I}$$
$$= \log k_4^0 - 2A\sqrt{I}$$

where the Debye–Hückel constant $A = 0\cdot5091$ at 25°C. Hence

$$\log \frac{k_4 k_2}{k_3} = \log \frac{k_4^0 k_2}{k_3} - 2A\sqrt{I}$$

The Brønsted Primary Salt Effect

B. D. COSTLEY,

Wolverhampton College of Technology

Discussion

According to the transition state theory of reaction rates the reactants are in equilibrium with an activated complex which may either dissociate into the reactants or form the products. The reaction may be represented by the equation

$$A + B \rightleftharpoons C\ddagger \rightarrow \text{Products}$$

where $C\ddagger$ represents the activated complex.

The rate of the reaction is assumed to be proportional to the concentration of the activated complex and the theory leads to the equation

$$\log k = \log k_0 + \log \frac{\gamma_A \gamma_B}{\gamma_{C\ddagger}} \tag{1}$$

where k is the velocity constant at a particular ionic strength,

k_0 is the velocity constant at zero ionic strength, and

γ represents the activity coefficient of each species.

The activity coefficients are related to the ionic strength of the solution by the Debye–Hückel equation or one of its extensions. For 1:1 electrolytes the form due to Güntelberg[1] is satisfactory for ionic strengths up to 0·1. Thus at 25°C

$$-\log \gamma_i = 0{\cdot}509 z_i^2 \frac{\mu^{1/2}}{1 + \mu^{1/2}} \tag{2}$$

for the activity coefficient of the ith kind of ion when μ is the ionic strength and z_i the ionic valency.

Combining equations (1) and (2) leads to the relationship

$$\log k = \log k_0 + 1{\cdot}018 z_A z_B \frac{\mu^{1/2}}{1 + \mu^{1/2}} \tag{3}$$

By determining the velocity constants at different ionic strengths a graph of $\log k$ against $\mu^{1/2}/(1+\mu^{1/2})$ can be plotted. This should be linear and of slope $1 \cdot 018 z_A z_B$.

The reaction chosen for study is that between the dye malachite green and the hydroxyl ion. The most significant resonance forms of malachite green are given by structures I and II. The contribution of form II suggests that there will be an electron deficiency at the tertiary carbon atom and this point is vulnerable to attack by hydroxyl ion to yield the colourless carbinol III.

I
(Quinonoid form)

II
(Benzenoid form)

III

The reaction rate can be followed spectrophotometrically.

In the presence of excess hydroxyl ion the reaction rate follows first-order kinetics and will thus obey the equation

$$t = \frac{2 \cdot 303}{k^1} \log \frac{[\text{Dye}_0]}{[\text{Dye}]} \tag{4}$$

where $[\text{Dye}_0]$ is the concentration of malachite green at $t = 0$
and $[\text{Dye}]$ is the concentration of malachite green at $t = t$.

k^1 is the first-order rate constant.

The concentration of dye is proportional to the absorbance at time t since Beer's law is obeyed. Thus, if A_0 and A are the absorbances of the solution at time $t = 0$ and $t = t$, respectively, equation (4) may be written

$$t = \frac{2 \cdot 303}{k^1} \log A_0 - \frac{2 \cdot 303}{k^1} \log A \tag{5}$$

A graph of t against $\log A$ will thus be linear of slope $-2 \cdot 303/k^1$. The second-order rate constant may be found from the relation

$$k^1 = k[\text{OH}^-] \tag{6}$$

where k is the second-order rate constant and $[\text{OH}^-]$ is the hydroxyl ion concentration which is virtually constant under the conditions used.

Apparatus and Chemicals

Spectrophotometer, e.g. Unicam SP.600 with thermostatted cell holder, six 100 ml graduated flasks, stop-clock, $1 \cdot 0$ M potassium nitrate solution, $0 \cdot 1$ M sodium hydroxide solution, malachite green.

Method

A stock dye solution is prepared by dissolving about 25 mg of malachite green in distilled water and diluting to 1 l.

Twenty millilitres of this solution is pipetted into a 100 ml graduated flask. Distilled water is added until the volume of the solution is about 95 ml. The solution is mixed, 5 ml of $0 \cdot 1$ M sodium hydroxide solution is added and the stop-clock started when half the caustic soda solution has run from the pipette. The volume is quickly adjusted to the mark, the solution mixed and a small amount is transferred to a 1 cm cell. The absorbance at 613 mμ is read at 30 sec intervals for 4 min then at 1 min intervals for a further 4 min. It is best to note the time at the instant the instrument is balanced and then to read the absorbance afterwards.

The procedure is repeated with five more solutions containing 2, 4, 6, 8 and 10 ml, respectively, of $1 \cdot 0$ M potassium nitrate solution in addition to the reactants.

Graphs are plotted of time against $\log A$ for each run and from the slopes of these, k^1 and k are calculated. Finally $\log k$ against $\mu^{1/2}/(1 + \mu^{1/2})$ is plotted and the slope compared with the value obtained from equation (3).

By investigating the reaction with other dyes of the triphenylmethane group, the velocity constants should be correlated with the amount of benzenoid form contributing to the resonance structure.

[1] E. GÜNTELBERG, *Z. Phys. Chem.*, **123**, 199 (1926).

The Variation of Rate Constant with Catalyst Concentration by a Polarimetric Method

Discussion

Experiment 70 describes a method for determining the forward velocity constant k_1 for the hydrolysis of 1-menthyl formate. If this experiment is carried out at 25°C and 35°C and at a variety of acid concentrations, it can be shown that the velocity constants k_{25} and k_{35} at these two temperatures are related to the acid concentration by the following equations:

$$k_{25} = k(\text{HCl}) \tag{1}$$

$$k_{35} = k'(\text{HCl}) \tag{2}$$

where k and k' are constants.

The energy of activation E for this hydrolysis as a function of acid concentration can then be determined by substituting equations (1) and (2) into the integrated form of the Arrhenius equation

$$\log \frac{k_2}{k_1} = \frac{E(T_2 - T_1)}{2 \cdot 303 R T_1 T_2}$$

where k_1 and k_2 are velocity constants at temperatures T_1 and T_2, respectively.

The equilibrium constant K for this reaction is given by the expression (equation (2), Experiment 70)

$$K = \frac{x_e^2}{(2 - x_e)(b - x_e)}$$

The K values for a number of acid concentrations at two different temperatures may be determined. The variation of heat of reaction ΔH with acid concentration can then be investigated by application of the integrated form of the van't Hoff equation

$$\log \frac{K_2}{K_1} = \frac{\Delta H(T_2 - T_1)}{2 \cdot 303 R T_2 T_1}$$

Apparatus and Chemicals

As for Experiment 70.

Method

The rate constant is determined as in Experiment 70 at acid concentrations of 0·1 M, 0·2 M and 0·4 M. The density of each equilibrium solution is measured. The temperature of the polarimeter water jacket is raised to 35°C and the whole procedure repeated.

The Measurement of Surface Area by the B.E.T. Method

M. J. JAYCOCK,

Loughborough University of Technology

Discussion

The specific surface area of a powdered sample, that is the surface area per gram of material, is a frequently determined parameter during the characterization of a powder. Of the many methods employed, that due to Brunauer, Emmet and Teller[1,2] is one of the most popular. In principle the method is comparatively simple. If the amount of gas required to cover the surface of a particular sample with a complete monolayer of adsorbate can be evaluated, then the surface area can be calculated if the area occupied by a single adsorbate molecule is known.

The B.E.T. equation is a theoretical description of the adsorption isotherm of a gas on a solid, and describes how the amount of gas adsorbed, expressed as an equivalent volume at S.T.P., v, varies with the equilibrium pressure, p. It has the form

$$\frac{p}{v(p_0-p)} = \frac{1}{v_m C} + \frac{C-1}{v_m C} \cdot \frac{p}{p_0}$$

where p_0 is the saturation vapour pressure of the adsorbate, v_m is the equivalent volume of gas at S.T.P., in a complete monolayer, and C is a constant characteristic of the particular system. A straight line results if $p/v(p_0-p)$ is plotted against p/p_0, the relative pressure. The value of v_m can be calculated from the slope, $(C-1)/v_m C$, and the intercept, $1/v_m C$, of this graph. Thus $v_m = 1/(\text{slope}+\text{intercept})$. The constant C is related to the heat of adsorption of the first layer, E_1, by

$$C = \exp\frac{E_1-E_L}{RT}$$

where E_L is the heat of liquefaction of the bulk liquid adsorbate. Most experimental systems may be described by the B.E.T. equation over the relative pressure range 0·05 to 0·35[3].

The most commonly used adsorbate is nitrogen at liquid nitrogen temperatutes, $77 \cdot 5°K$, and at this temperature the area occupied by a single nitrogen molecule in a complete monolayer is $16 \cdot 2 Å^2$. If pure liquid nitrogen is used as the thermostat liquid surrounding the sample tube, then the saturation vapour pressure of nitrogen at the adsorption temperature will be equal to the atmospheric pressure.

Apparatus and Chemicals

The experimental apparatus is based on a design by Krieger,[4] but has a larger volume in the gas burette and a reduced "dead space" above the sample. A line diagram shows the arrangement of the components of the vacuum system. The volumes of the burette bulbs are

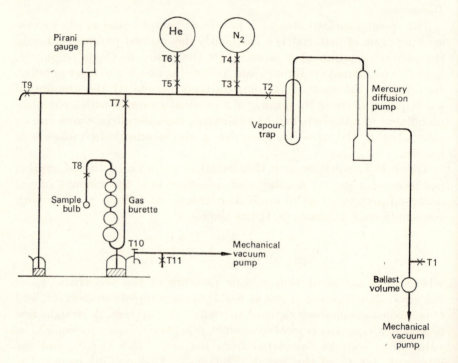

FIG. 1. Line diagram of adsorption apparatus.

approximately 5, 15, 25, 50 and 125 ml and these volumes will have been accurately determined before the apparatus was assembled. T9 is not used during the investigation; it is used to fill the helium and nitrogen storage vessels from cylinders of the gases.

The amount of sample chosen (carbon black, alumina, silica, etc.) should have a total surface of about 20 m². It is possible to work with higher and lower values of surface area but the figure quoted for the initial pressure of nitrogen in the following section will need adjustment.

Method

Evacuating the system

Taps T1, T2, T10 and T11 are closed and both mechanical vacuum pumps are switched on. The water supply to the mercury diffusion pump is turned on and the heater is switched on. The Dewar flask around the vapour trap is filled with liquid nitrogen and topped up from time to time as necessary.

Taps T4, T6, T7 and T9 are closed and taps T3 and T5 are opened. Tap T2 is slowly opened. The pressure changes in the system are followed with a Pirani gauge until the pressure is less than 10^{-3} torr. (If the system does not seem to be pumping out very readily a demonstrator should be consulted since there may be a leak in the system.) Taps T3, T5 and T8 are closed.

T7 is partially opened very slowly and the mercury level is raised to the bottom of the bend in the gas burette when T7 is closed again. T10 is slowly opened to the vacuum line until the mercury level is lowered to its original position when T10 is again closed. The operations with T7 and T10 are repeated until there is no further movement of the mercury level. T7 is then opened and the pressure changes in the system are followed with the Pirani gauge. The system should pump down to a pressure of less than 10^{-3} torr fairly quickly.

A weighed sample of the powder whose surface area is to be measured is placed in the sample tube which is illustrated in Fig. 2. The filler rod is replaced, the joint greased and connected to the appropriate socket on the frame. T8 is opened very very slowly. If this is not done very carefully, some of the sample will be sucked out of the sample tube. The system is allowed to pump out until a pressure lower than 10^{-3} torr can be maintained with T2 shut. This stage can be reached more rapidly by placing a small furnace round the bulb of the sample tube and heating it to a suitable temperature, which will be determined either by the nature of the sample or by the softening temperature of the glass.[5] T8 is closed.

Volume calibration with helium

T2 is closed. T6 is opened and the space between T5 and T6 is filled with helium when T6 is closed. With T7 open and T8 shut, T5 is opened and helium is admitted to the gas burette when T5 is closed again. The manometer attached to the system will enable the pressure of helium in the system to be estimated and this should be approximately 20 torr. If the pressure of helium is not sufficiently great more

M

helium should be admitted by repeating the procedure described above with taps T5 and T6. If the pressure of helium in the system is too great, then some of the gas may be pumped out by partially opening T2 until a suitable pressure remains.

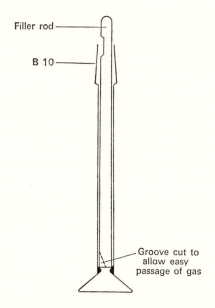

Fig. 2. The sample tube.

Sufficient air is admitted through T10 to raise the mercury level in the gas burette so as to partially fill the bottom bulb. T2 is opened and the part of the system above the manometer limb of the gas burette is evacuated. The pressure of helium is then equivalent to the difference in height of the mercury levels in the two limbs of the gas burette. The mercury level is raised by means of T10 and adjusted to each of the top four calibration marks of the gas burette in turn, the pressure being recorded in each case. Since the quantity of gas has been kept constant during these measurements, the ideal gas law may be used to determine graphically the volume between the upper calibration mark and T8.

A Dewar flask, kept full of liquid nitrogen, is placed around the sample tube, such that the liquid nitrogen level is kept at some fixed mark near the middle of the stem of the sample tube. Physical adsorption only occurs to an appreciable extent at pressures and temperatures close to those required for liquefaction. Helium will not thus be measurably adsorbed at liquid nitrogen temperatures and consequently may be used for volume calibrations at this low temperature. The pressure measurements at the top four calibration marks of the gas burette are repeated and the equivalent volume of the sample tube, which is usually termed the "dead space," is determined graphically.

T7 is closed and the mercury level in the gas burette is lowered by using T10 until it is below the bend. T7 is opened and the system is evacuated until a pressure of less than 10^{-3} torr can be maintained with T2 shut.

Determination of nitrogen adsorption

With T8 shut nitrogen is admitted to the gas burette from the storage vessel in a similar manner to that employed with the helium until a pressure of about 100 torr is obtained. A sample of gas is trapped in the burette and the system above the manometer limb is pumped out. The pressure is measured with the mercury level at the bottom calibration mark. A knowledge of the pressure and the volume occupied by the nitrogen in the gas burette enables the quantity of gas present to be calculated by means of the ideal gas law.

With the Dewar vessel full of liquid nitrogen in exactly the same position around the sample tube as when the volume calibration was carried out, T8 is opened and the mercury level is kept adjusted to the bottom calibration mark. When the pressure is steady, its value is noted. The drop in pressure is due to two factors: the increase in volume on opening T8, and the adsorption of some of the nitrogen in the system. The pressure and volume of the system at equilibrium enable the quantity of nitrogen remaining unadsorbed in the system to be determined and the difference between this value and the quantity of nitrogen originally trapped in the gas burette represents the amount adsorbed by the sample at the equilibrium pressure.

The mercury level is raised to the next calibration mark, and the equilibrium pressure and the corresponding amount adsorbed again determined. This process is repeated for each of the remaining calibration marks.[6] The temperature of the air in the jacket surrounding the gas burette is noted and the atmospheric pressure is recorded.

T7 is closed and the mercury level is slowly lowered until it is below the bend in the burette. The Dewar vessel of liquid nitrogen is removed from the sample tube which is allowed to return to room temperature. T7 is opened slowly and the system is evacuated.

Closing down the system

All taps are set to the closed positions. The mercury diffusion pump is switched off and allowed to cool before the water supply is turned off. The mechanical vacuum pumps are switched off and T11 and T1 are immediately opened. The Dewar vessel of liquid nitrogen is removed from the vapour trap.

Treatment of the results

The results are in the form of pairs of values of the amount adsorbed and the corresponding equilibrium pressure. The amount adsorbed is converted to an equivalent volume at S.T.P., per gram of adsorbent and the B.E.T. equation is applied as described in the Discussion. The specific surface area[7] and the B.E.T. constant C are calculated.

Note. If the investigation is to be in everyday use, then the system should not be closed down each day, but left evacuating overnight with a sample permanently attached. In this case, the sections *Evacuating the system* and *Closing down the system* should be ignored.

[1] S. BRUNAUER, P. H. EMMETT and E. TELLER, *J. Amer. Chem. Soc.*, **60,** 309 (1938); for errata see P. H. EMMETT and T. W. DE WITT, *Industr. Eng. Chem.* (anal.), **13,** 28 (1941).

[2] D. M. YOUNG and A. D. CROWELL, *Physical Adsorption of Gases*, Butterworths, 1962, p. 147.

[3] Ref. 2, p. 190.

[4] K. A. KRIEGER, *Industr. Eng. Chem.* (anal.), **16,** 398 (1944).

[5] Ref. 2, pp. 277–9.

[6] Ref. 2, p. 279.

[7] Ref. 2, p. 182.

Critical Micelle Concentration

M. J. JAYCOCK,
Loughborough University of Technology

Discussion

Ionic surface active materials, such as sodium dodecylsulphate (SDS), consist of two species of ion, an ionic head group attached to a hydrocarbon tail (dodecyl-sulphate ion) and a counter ion (sodium ion). The way in which the physical properties of a solution of a typical colloidal electrolyte, such as SDS, vary with increasing concentration, alters when a particular concentration is reached. This concentration is known as the critical micelle concentration, or c.m.c. The accepted explanation of this is that in the region of the c.m.c., aggregation of the long chain electrolyte ions into quite large, charged units begins to occur.[1,2] These charged units are typically made up of about 100 surface active ions forming a spherical unit, with the charges situated at the periphery and the hydrocarbon tails in the centre. The adsorption of a considerable number of counterions reduces the overall charge and also the coulombic repulsion between the head groups.

Apparatus and Chemicals

Conductance bridge, conductivity cell, thermostat, Du Nony tensiometer[3] and density bottle. Sodium dodecyl-sulphate B.P. and a purified sample of the same material. (Impurities such as the hydrolysis products

$$NaDSO_4 + H_2O \rightarrow NaHSO_4 + DOH$$

have been partially removed by recrystallization from water at 0°C and extraction with petroleum ether.)

Method

A 2×10^{-4} M aqueous solution of the purified material is made up. This solution should be clear if the material is sufficiently pure. All other concentrations may be made by dilution of this solution. The conductance, surface tension and density of solutions are measured

343

over the concentration range from 10^{-3} M to 2×10^{-2} M. The surface tension of solutions of B.P. grade material is measured over the same concentration range.

Tabulate the values of the c.m.c. obtained by the various methods and comment fully on these results. Discuss the significance of the surface tension results obtained with the B.P. grade material.

[1] K. Shinoda, T. Nakagawa, B. Tamamushi and T. Isemura, *Colloidal Surfactants*, Academic Press (1963), Chapter I.

[2] A. W. Adamson, *Physical Chemistry of Surfaces*, Interscience (1960), pp. 373–8.

[3] J. T. Davies and E. K. Rideal, *Interfacial Phenomena*, Academic Press. 2nd ed. (1963), p. 43.

The Protonation of Aldehydes and Ketones in Sulphuric Acid Media

Discussion

In a strong protogenic solvent, e.g. aqueous sulphuric acid, aldehydes and ketones behave as very weak bases. The organic molecule accepts a proton from the solvent to give an oxonium salt and the equilibrium is

$$\underset{R}{\overset{R}{\diagdown}}C = \overset{+}{O}H + HSO_4^- \rightleftharpoons \underset{R}{\overset{R}{\diagdown}}C = O + H_2SO_4$$

The measurement of the extent of the ionization and the equilibrium constant K_a for the dissociation of the conjugate acid $R_2C = O^+H$ was originally made by Hammett[1] using a spectrophotometric method. Extensive pK_a values for weak bases of this type have more recently been made available by the work of Davis and Geissman,[2] who measured the basicity of substituted flavones, and by Stewart et al.[3] who investigated some aromatic aldehydes and ketones.

The calculation of the pK_a values from spectral data was carried out by Davis and Geissman as follows. Let

α = fraction of the compound in the protonated form.

ϵ_p = the molar absorptivity of the protonated form,

ϵ_u = the molar absorptivity of the unprotonated form,

λ_u = a selected wavelength where $\epsilon_u \gg \epsilon_p$,

λ_p = a selected wavelength where $\epsilon_p \gg \epsilon_u$,

ϵ = the observed molar absorptivity of a mixture of the protonated and unprotonated forms,

then if Beer's Law is valid for the experimental conditions

$$(\epsilon)\lambda_p = [(\epsilon_u)\lambda_p + (\epsilon_p)\lambda_p - (\epsilon_u)\lambda_p]\alpha \tag{1}$$

$$(\epsilon)\lambda_u = [(\epsilon_p)\lambda_u + (\epsilon_u)\lambda_u - (\epsilon_p)\lambda_u]1 - \alpha \tag{2}$$

On subtracting (2) from (1) and letting

$$y = (\epsilon)\lambda_p - (\epsilon)\lambda_u \tag{3}$$

we obtain

$$y = [(\epsilon_u)\lambda_p - (\epsilon_u)\lambda_u] + [(\epsilon_p)\lambda_p - (\epsilon_u)\lambda_p + (\epsilon_u)\lambda_u - (\epsilon_p)\lambda_u]\alpha \qquad (4)$$

Since values such as $(\epsilon_u)\lambda_p$ may be assumed to be approximately constant over a small range of acid concentrations in the neighbourhood of 50 per cent protonation, then the fraction protonated, α is linearly related to y, i.e.

$$\alpha = my + k \qquad (5)$$

where

$$m = 1/[(\epsilon_p)\lambda_p - (\epsilon_u)\lambda_p + (\epsilon_u)\lambda_u - (\epsilon_p)\lambda_u] \qquad (6)$$

and

$$k = [(\epsilon_u)\lambda_u - (\epsilon_u)\lambda_p]/[(\epsilon_p)\lambda_p - (\epsilon_u)\lambda_p + (\epsilon_u)\lambda_u - (\epsilon_p)\lambda_u] \qquad (7)$$

In his original investigations Hammett[4] defined a function which expresses the tendency of a solvent to transfer a proton to a weak base in the same way as the pK_a value measures the tendency of the base to accept the proton. Hammett called this function the acidity function H_0,

$$H_0 = -\log\frac{a_{H^+} \cdot \gamma_B}{\gamma_{BH^+}} \qquad (8)$$

where a_{H^+} is the activity of the hydrogen ion, γ_B is the activity coefficient of the weak base, and γ_{BH^+} is the activity coefficient of the conjugate acid. The pK_a of the conjugate acid is given by

$$pK_a = -\log\frac{a_{H^+} \cdot a_B}{a_{BH^+}} \qquad (9)$$

$$= -\log\frac{a_{H^+}c_B\gamma_B}{c_{BH^+}\gamma_{BH^+}} \qquad (10)$$

where c is the concentration. Hence on combining equation (8) with equation (10) we obtain

$$H_0 = pK_a + \log(c_B/c_{BH^+}) \qquad (11)$$

$$= pK_a + \log([unprotonated]/[protonated]) \qquad (12)$$

But since

$$[unprotonated]/[protonated] = [1 - my - k]/[my + k] \qquad (13)$$

$$H_0 = pK_a + 0.434\ln[1 - my - k]/[my + k] \qquad (14)$$

The slope of the curve obtained by plotting H_0 vs. y is dH_0/dy and at the point of inflection

$$d^2H_0/dy^2 = 0 = 0.434c^2[1/(1 - my - k)^2 - 1/(my + k)^2] \qquad (15)$$

At this point of inflection equation (15) gives

$$1 - (my + k) = (my + k) \tag{16}$$

and from equation (14) and (15)

$$H_0 = pK_a \tag{17}$$

Apparatus and Chemicals

Hilger and Watts Uvispek spectrophotometer, Unicam S.P.800 spectrophotometer, Adkins Thermostatted Cell Holder, ten 50 ml graduated flasks, sulphuric acid (M.A.R. grade), acetophenone, and 2·5 N sodium hydroxide.

Method

Sulphuric acid–water solutions ranging from 40·0 to 98·0 per cent w/w acid are prepared and the concentration of the acid determined by titration with the standard base. The H_0 values of these solutions are obtained from Table 1 (data from Paul and Long).[5]

TABLE 1

H_2SO_4 % w/w	$-H_0$	H_2SO_4 % w/w	$-H_0$
40	2·41	80	6·97
45	2·85	85	7·66
50	3·38	90	8·27
55	3·91	95	8·86
60	4·46	96	8·98
65	5·04	97	9·14
70	5·65	98	9·36
75	6·30		

A sample of acetophenone is weighed out on a micro-balance and dissolved in acetone to give a stock solution of approximately 5×10^{-4} M. With the aid of a pipette, 5 ml portions of the solution are added to the graduated flasks. The acetone is removed with a pump and the residual ketone is dissolved in the sulphuric acid–water solutions. The flasks are placed in a thermostat at 25°C.

The ultraviolet spectra of acetophenone in 40 per cent and 98 per cent aqueous sulphuric acid are measured using the SP.800 spectrophotometer. The wavelengths of maximum absorption of the protonated (λ_p) and the unprotonated (λ_u) species are noted. The molar absorptivity of the ketone in aqueous sulphuric acid solutions is measured at these two wavelengths, in a thermostatted cell using the Hilger–Uvispek spectrophotometer. A plot of ($y \times 10^{-3}$) vs. H_0 is prepared and the pK_a value of the acetophenone is determined.

pK_a values of other aldehydes or ketones may be determined in a similar way and for a related series of carbonyl compounds the pK_a values should be discussed in relation to the electronic or steric effects of the substitutents.

[1] L. A. FLEXSER, L. P. HAMMETT and A. DINGWALL, *J.A.C.S.*, **57**, 2103 (1935).

[2] C. T. DAVIS and T. A. GEISSMAN, *J.A.C.S.*, **76**, 3507 (1954).

[3] R. STEWART and K. YATES, *J.A.C.S.*, **80**, 6355 (1958); *Can. J. Chem.*, **37**, 664 (1959); R. STEWART, M. GRANGER, R. MOODIE and L. MUENSTER, *Ibid.*, **41**, 1605 (1963).

[4] L. P. HAMMETT, *Physical Organic Chemistry*, McGraw-Hill, 1940.

[5] M. A. PAUL and F. A. LONG, *Chem. Rev.*, **57**, 1 (1957).

The Anodic Behaviour of Metals

N. A. HAMPSON,

Loughborough University of Technology

Discussion[1,2]

The reaction

$$Me \rightarrow Me^{++} + 2e$$

may occur at an anode. The ions produced may be hydrated species but are more likely to be intermediates of low charge and coordination number since these pass more readily through the double layer.

If the applied potential exceeds a certain critical limit, passivation occurs and the dissolution process cannot occur at the surface of the electrode which becomes blocked by a film of insoluble basic salt, usually the hydroxide or oxide. Under these conditions the potential rises to that required for oxygen evolution.

The ability of a system to withstand passivation is connected with the capacity of the system for the anodic product. Under conditions of pure diffusion the time required for the concentration of the anodic product in the solution immediately adjacent to the electrode can be found by solving the diffusion equation

$$\frac{\partial c}{\partial t} = \frac{\partial^2 c}{\partial x^2}D \tag{1}$$

where c is the concentration at a distance x from the electrode and D is the diffusion coefficient of the anodic product. The solution of equation (1) leads to an expression of the form

$$i = kt^{-1/2} \tag{2}$$

where i is the current density, t is the time to passivate and k is a constant for the system.

In practice, most systems are generally more complex than the simple case considered as diffusion is not the only mode of mass transfer involved. Convection becomes extremely important, particularly with vertically oriented anodes, e.g., plating baths, electrorefining systems, etc. Under such conditions equation (2) is modified to

$$(i - i_l) = kt^{-1/2} \tag{3}$$

where i_l is the limiting current density below which passivation does not occur. For experiments with passivation times of less than 1000 sec, equation (2) applies if horizontally oriented anodes are used.

Apparatus and Chemicals

Zinc sheet, carbon rod 1 cm in diameter, electrolytic cell, valve voltmeter, resistor, 30 volt d.c. supply, switch, ammeter, clips, rubber ring, 1 M, 2 M, 3 M, 4 M and 5 M solutions of potassium hydroxide, carbon tetrachloride, stop-watch.

A suitable cell may be made from a 2 cm diameter pyrex tube. One end of this tube should be ground flat. The zinc sheet should fit under the rubber ring and the ground face of the tube presses on the sheet.

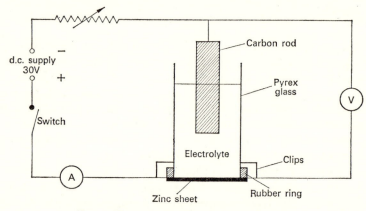

FIG. 1. Electrolytic cell and circuit diagram.

Method

The zinc sheet is cleaned with carbon tetrachloride.

The electrolytic cell is mounted vertically so that the zinc anode is horizontally oriented. The cathode is a carbon rod. A steady current (current density 0·01 to 0·2 A cm^{-2}) is applied to the cell and the voltage across the electrodes is measured with a valve voltmeter. The passivation time is measured from the time of closing the circuit to the time the anode potential rises steeply as indicated by a sharp increase in the voltage across the cell. This should coincide with a film flashing across the anode.

The passivation time is a function of current density and the results may be correlated in terms of equations (2) and (3). The amount of mass transfer by modes other than diffusion should be assessed. The effects of changes of electrolyte concentration should be investigated.

[1] N. A. HAMPSON and M. J. TARBOX, *J. Electrochem. Soc.*, **110**, 95 (1963).
[2] N. A. HAMPSON, M. J. TARBOX, J. T. LILLEY and J. P. G. FARR, *Electrochem. Tech.*, **2**, 309 (1964).

The Polarograph: A Study of the Variables when Interpreting Polarograms

Discussion

The polarographic method is now a well-established analytical technique in that under certain conditions the diffusion current (i) is linearly proportional to the concentration of the reducible species (c). The Ilkovic equation relating instantaneous current values (i) and concentration (c) is given in Experiment 85. It is worth investigating, however, various factors which can influence the linear property of the current concentration relationship. The normal polarogram as obtained in Experiment 85 occurs when the following conditions prevail:

(a) The cell solution is completely deaerated.
(b) The concentration of the potassium chloride supporting electrolyte is at least ten times greater than the concentration of the species under test.
(c) The temperature of the cell is constant.
(d) One mercury drop forms over 3 to 5 sec.
(e) The minimum concentration of maximum suppressor (gelatine) is added to the cell solution.

A quantitative study of these five factors can be made.

Apparatus and Chemicals

Cambridge automatic recording polarograph, cylinder of oxygen-free nitrogen, 10^{-4} solutions of lead and zinc ions, potassium chloride and gelatine.

Method

An ordinary polarographic analysis is performed using a mixture of 10^{-4} M zinc solution and 10^{-4} M lead solution as described in Experiment 85. The variables in this technique as listed above are altered consecutively.

(a) The solution is deaerated by bubbling nitrogen gas for periods of 5 min, 10 min, 15 min, 20 min and 25 min and a polarogram is recorded after each deaeration time. These bubbling times may be greater or smaller than the values quoted and are dependent upon the volume of solution under test and the bubbling rate of the nitrogen.

(b) Six polarograms are obtained using potassium chloride as the supporting electrolyte at concentrations of 0·0005 M, 0·00075 M, 0·001 M, 0·0015 M, 0·0020 M and 0·0025 M.

(c) Three polarograms should be recorded at 25°C, 35°C and 45°C.

(d) The five polarograms for mercury dropping rates of 1 sec, 2 sec, 4 sec, 6 sec and 8 sec are recorded.

(e) Finally the seven polarograms are recorded when progressive additions of a maximum suppressor (gelatine) are in the test solution (i.e. 0·01 per cent, 0·05 per cent, 0·1 per cent, 0·5 per cent, 1 per cent, 5 per cent and 10 per cent concentrations of gelatine).

The polarograms obtained under various experimental conditions are interpreted and comment should be made on the constancy or otherwise of the half wave potential and the linear relationships between the limiting current and the concentration of the reducible species.[1]

[1] For a further theoretical background to this investigation see O. H. MULLER, *The Polarographic Method of Analysis,* Chemical Educ. Publ. Co. (1956).

Differential Potentiometric Titrations

Discussion

Potentiometric titrations have already been described in Experiment 50 when the e.m.f. of a given electrode assembly was recorded throughout the titration. To obtain an accurate end point $\Delta E/\Delta v$ values were plotted against v as shown in Fig. 1. It is possible, however, by constructing a suitable apparatus, to directly record the differential e.m.f. value ΔE. This may be done with a duplicate electrode assembly, with which the small voltage difference ΔE is measured as the titration proceeds.

Apparatus and Chemicals

Tinsley potentiometer, stirrer, silver electrodes, 0·1 M silver nitrate solution, and 0·1 M potassium chloride solution.

Method

A convenient study is the titration of silver nitrate with potassium chloride, using a duplicate silver electrode assembly. The titration cell assembly is prepared as shown in Fig. 2. A and B are silver electrodes immersed in the silver nitrate solution, the electrode B is enclosed in a glass jacket with a small orifice at one end. Initially the solution around each silver electrode will contain silver ions of the same activity and there will be no difference of potential between the electrodes.

A small amount, Δv ml, of potassium chloride solution is added from the burette C to the silver nitrate solution. The solution surrounding electrode A will undergo a decrease in silver ion activity and hence its potential will decrease. The potential of electrode B will, however, remain as before because the solution surrounding this electrode is isolated from the bulk of the solution, with the result that there will have been no change in the silver ion activity. A difference of potential ΔE will thus exist between the two electrodes.

The solution around electrode B is then expelled by means of the rubber teat D. On refilling this compartment all of the solution in the system will be homogeneous and the difference of potential between the electrodes will again be zero. This procedure is then repeated by adding a further small volume, Δv ml of potassium chloride solution, and the ΔE value measured. This process is continued until the titration is complete and the full differential curve obtained. The initial addition of the potassium chloride can be quite rapid as the end point will be apparent from the increasing ΔE values.

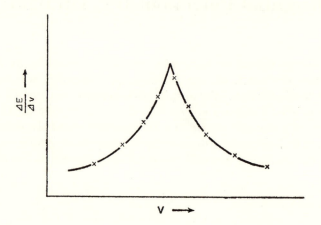

FIG. 1. Variation of the potentiometric differential with added volume of titrant.

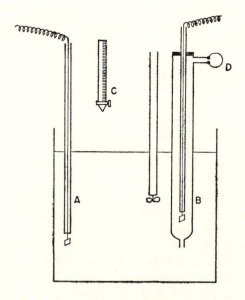

FIG. 2. The titration cell.

This apparatus must be designed so that the orifice in the glass shield around electrode B is of critical size to permit a fairly rapid but thorough expulsion of the liquid but should not be too large to permit mixing of the solutions at other times.

The possibilities of this apparatus, using different electrode assemblies, should be investigated for other types of titration, e.g. platinum electrodes for acid alkali titrations.

Statistical Treatment of Experimental Data

N. FERGUSON,
Loughborough University of Technology

Types of Error

The measurement of physical magnitudes always contain errors. These errors may be divided into two classes:

(i) constant or systematic errors and

(ii) random or accidental errors.

Errors of the first class enter into observations through the use of faulty graduated scales or the non-application of some correction factor, for example, when one neglects a buoyancy correction in accurate weighings. It is up to the experimenter to look carefully for possible causes of systematic errors and to remove them or at least to make some standard or empirical correction to eliminate their effect. Constant errors can be particularly serious if one is not careful since they do not reveal themselves in repeat measurements. They can usually be detected when two or more quite distinct experimental procedures are used to measure the same quantity.

Random errors produce fluctuations in repeat measurements performed under apparently the same conditions. These errors are caused by slight changes in disturbing factors not under the experimenter's control. The value of the random error in a particular observation is not predictable, but the average of the random errors in a large series of observations is approximately zero.

The object of measuring a particular quantity is to determine as closely as possible its "true" value, which will be denoted by μ. In general, the measurement is repeated n times and observations

$$x_1, x_2, \ldots, x_i, \ldots, x_n$$

are obtained. x_i denotes the ith observation in the set. If we denote the constant error in each of the n observations by b and the random error in the ith observation by e_i, then

$$x_1 = \mu + b + e_1$$
$$x_2 = \mu + b + e_2$$
$$\cdot \quad \cdot \quad \cdot \quad \cdot \quad \cdot$$
$$x_i = \mu + b + e_i,$$
$$\cdot \quad \cdot \quad \cdot \quad \cdot \quad \cdot$$
$$x_n = \mu + b + e_n$$

On adding the n equations and dividing by n, we get

$$\frac{\sum\limits_{i=1}^{n} x_i}{n} = \frac{n\mu}{n} + \frac{nb}{n} + \frac{\sum\limits_{i=1}^{n} e_i}{n}$$

that is,

$$\bar{x} = \mu + b + \bar{e},$$

where

$$\bar{x} = \frac{\sum\limits_{i=1}^{n} x_i}{n} \quad \text{and} \quad \bar{e} = \frac{\sum\limits_{i=1}^{n} e_i}{n}$$

$\bar{x}$ and $\bar{e}$ being the arithmetic means of the observations and random errors respectively. It is now expedient to distinguish between accuracy and precision.

Accuracy and Precision

The accuracy of an observation is measured by the deviation of the observation from the true value, that is, by

$$x_i - \mu = b + e_i$$

in the case of the ith observation. The precision of an observation, on the other hand, is measured by the deviation of the observation from the average of all the observations, that is, by

$$x_i - \bar{x} = e_i - \bar{e} \simeq e_i$$

since as we have noted above $\bar{e} \simeq 0$. Thus high precision only implies great accuracy when there is no constant error, i.e. when $b = 0$. Clearly high accuracy always implies high precision.

Some Statistical Ideas and Methods

Statistical methods are designed to estimate the precision of experimental results. In order to determine the precision of a set of measurements it is essential to retain in the data the terminal decimal places which are often rounded off, since these are the ones that contain the information on the variation between replicate determinations. When a large number of replicate measurements are examined, it is seen that they fluctuate around the arithmetic mean of the set with high density close to the mean and rapidly decreasing density as the values deviate further from the mean. In many situations the distribution of the measurement is symmetrical about the mean and follows what is known as the normal distribution. Figure 1 shows the relative frequency of occurrence of values of a measurement which is normally distributed. C is the centre of the distribution, that is, the value about which all the observations are scattered. For measurements whose true values are μ, but in all of which there is a constant error b, the centre of the distribution is at $x = \mu + b$. This point is termed the population mean. Only when the constant error b has been removed from all the measurements is the population mean at the true value $x = \mu$. In what follows we shall suppose that this adjustment has been made.

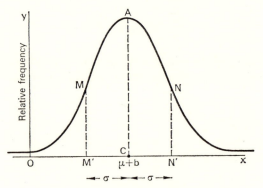

Fig. 1. The normal distribution

M and N, in the above figure, are the two points of inflection on the normal curve, and are situated at $x = \mu - \sigma$ and $x = \mu + \sigma$, where $\sigma = M'C = N'C$. σ^2 is called the population variance and σ the standard deviation of the population. Clearly σ measures, in some way, the width of the distribution since it is the distance of the points of inflection from the centre of the distribution.

The equation of a normal distribution whose mean is μ and whose variance is σ^2 is given by

$$y = \frac{1}{\sqrt{(2\pi)}\sigma} e^{-(x-\mu)^2/2\sigma^2}, \quad -\infty \leqslant x \leqslant \infty$$

The two parameters μ and σ^2 completely specify the distribution. The problem with which we are generally faced is to estimate μ and σ^2 from n random observations

$$x_1, x_2, \ldots, x_n$$

It can be shown that for a normal distribution the population mean μ is best estimated by the arithmetic mean of the sample, that is by

$$\bar{x} = \frac{x_1 + x_2 + \ldots + x_n}{n} = \frac{\sum\limits_{i=1}^{n} x_i}{n}$$

The variance σ^2 can be best estimated by

$$S^2(x) = \frac{\sum_{i=1}^{n}(x_i - \bar{x})^2}{n-1}$$

$\sum_{i=1}^{n}(x_i - \bar{x})^2$ is the sum of the squares of the deviations of the observations from the mean. $n-1$ is called the number of degrees of freedom in the sample of n observations.

Computationally it is more convenient to make use of the identity

$$\sum_{i=1}^{n}(x_i - \bar{x})^2 \equiv \sum_{i=1}^{n}x_i^2 - \frac{(\sum_{i=1}^{n}x_i)^2}{n}$$

in calculating $S^2(x)$. Thus the standard deviation σ is estimated by

$$S(x) = \sqrt{\left(\frac{\sum x^2 - (\sum x)^2/n}{n-1}\right)}$$

If we have a measurement x which is normally distributed with mean μ and variance σ^2 we can uniquely express the measurement in terms of the number of standard deviations u by which it differs from the mean. Thus

$$u = \frac{x - \mu}{\sigma},$$

where u is called the standardized normal deviate corresponding to x. It can be proved that the random variable u is normally distributed with mean zero and variance unity, as shown in Fig. 2.

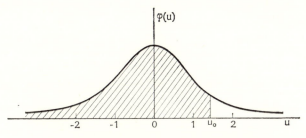

FIG. 2. The standardized normal distribution

The equation of the standardized normal distribution is given by

$$\phi(u) = \frac{1}{\sqrt{(2\pi)}} e^{-u^2/2}$$

where $-\infty < u < +\infty$.

The area to the left of the ordinate at $u = u_0$ is denoted by

$$\Phi(u_0) = \int_{-\infty}^{u_0} \phi(u)\, du = \int_{-\infty}^{u_0} \frac{1}{\sqrt{(2\pi)}} e^{-u^2/2}\, du$$

The value of $\Phi(u_0)$ gives the probability that u is less than or equal to u_0 in a standardized normal distribution or what is more useful the probability that the measurement x is less than or equal to $\mu + u_0\sigma$, i.e. the mean plus u_0 standard deviations. An extract from the table of $\Phi(u)$ is given in Table 1.

TABLE 1

u	$\Phi(u)$
1·00	0·8413
1·28	0·9000
1·64	0·9500
1·96	0·9750
2·33	0·9900
2·58	0·9950
3·09	0·9990
3·29	0·9995

From the symmetry of the Normal curve and the fact that the total area under the curve is unity

$$\Phi(-u) = 1 - \Phi(u)$$

When the standard deviation σ is accurately known we can use the above table to make statements such as:

$$\text{Prob}\,(-1{\cdot}96 \leqslant u \leqslant +1{\cdot}96) = \text{Prob}\,(\mu - 1{\cdot}96\sigma \leqslant x \leqslant \mu + 1{\cdot}96\sigma)$$
$$= 0{\cdot}95$$

i.e. on average 95 per cent of the observations should be in the range $\mu - 1{\cdot}96\sigma$ to $\mu + 1{\cdot}96\sigma$, or only 1 in 20 of the observations should fall outside these limits.

$$\text{Prob}\,(u > 1{\cdot}64) = \text{Prob}\,(x > \mu + 1{\cdot}64\sigma)$$
$$= 1 - 0{\cdot}95 = 0{\cdot}05$$

This statement means that on average only 1 observation in 20 should exceed $\mu + 1{\cdot}64\sigma$.

$$\text{Prob}\,(u < -3{\cdot}09) = \text{Prob}\,(x < \mu - 3{\cdot}09\sigma)$$
$$= 1 - 0{\cdot}9990 = 0{\cdot}001$$

This means that on average only 1 observation in 1000 should be less than $\mu - 3 \cdot 09\sigma$. Spurious observations can often be detected by checking how many standard deviations an anomolous-looking reading is away from the mean. Since only 1 observation in 1000 should exceed $\mu + 3 \cdot 09\sigma$ or be less than $\mu - 3 \cdot 09\sigma$, this gives a useful criterion.

Our interest does not usually lie in individual determinations but in the average of n determinations. We frequently need to know how precisely we have determined the true value on the basis of the arithmetic mean $\bar{x}$. This leads us to consider the distribution of the mean $\bar{x}$. When the individual observations x are normally distributed with mean μ and variance σ^2 the arithmetic mean $\bar{x}$ of n determinations is also normally distributed about the same mean μ but with variance σ^2/n. The standard deviation of the sample mean $\bar{x}$ is thus $\sigma/\sqrt{n}$. To place 95 per cent confidence limits on the true mean when the standard deviation σ is known, we compute

$$\bar{x} \pm 1 \cdot 96 \frac{\sigma}{\sqrt{n}}$$

Using these limits, we shall be correct in saying that the true mean μ lies within the range $\bar{x} \pm 1 \cdot 96\sigma/\sqrt{n}$ in approximately 95 such statements out of a 100. Similarly 99 per cent confidence limits for μ are given by

$$\bar{x} \pm 2 \cdot 58 \frac{\sigma}{\sqrt{n}}$$

the risk of the true mean μ being outside these limits being 1 per cent.

Tests Based on the t-distribution

In practice we seldom know σ sufficiently accurately to use the normal deviates u in making confidence statements. The problem with which we are often faced is, from observations

$$x_1, x_2, \ldots, x_n$$

where n is some integer, between perhaps 2 and 10, to place confidence limits on the true mean μ. This can be done by means of the t-distribution. When σ is replaced by an estimate $S(x)$ in $(\bar{x} - \mu)/(\sigma/\sqrt{n})$ it becomes $(\bar{x} - \mu)/(S/\sqrt{n})$. We see that there is now uncertainty in the denominator of this expression, since S is based on the n observations which contain random errors. It can be shown that $(\bar{x} - \mu)/(S/\sqrt{n})$ has a t-distribution with $(n-1)$ degrees of freedom when the observations are themselves a random sample from a normal distribution. Various percentage points of t have been tabulated for different numbers of degrees of freedom. An extract from the t-table is given in Table 2.

TABLE 2. PERCENTAGE POINTS OF THE t-DISTRIBUTION

No. of degrees of freedom, v	Probability of a deviation greater than t				
	0·005	0·01	0·025	0·05	0·1
1	63·66	31·82	12·71	6·31	3·08
3	5·84	4·54	3·18	2·35	1·64
5	4·03	3·36	2·57	2·01	1·48
10	3·17	2·76	2·23	1·81	1·37
20	2·85	2·53	2·09	1·72	1·32
120	2·62	2·36	1·98	1·66	1·29
∞	2·58	2·33	1·96	1·64	1·28

When we have n observations

$$x_1, x_2, \ldots, x_n;$$

which are normally distributed, we must compute the arithmetic mean

$$\bar{x} = \frac{\sum x}{n}$$

and the sample variance

$$S^2 = \frac{\sum x^2 - \dfrac{(\sum x)^2}{n}}{n-1}$$

and hence the standard error

$$S(\bar{x}) = \sqrt{\frac{S^2}{n}}$$

The deviation of the sample mean $\bar{x}$ from the true mean μ is then expressed in terms of the computed standard error $S(\bar{x})$, giving a t-factor based on $(n-1)$ degrees of freedom, i.e.

$$t_{n-1} = \frac{\bar{x} - \mu}{S(\bar{x})}$$

The above table may be used for making statements such as:

Prob. $[t_5 = (\bar{x} - \mu)/S(\bar{x}) > 2\cdot01] = 0\cdot05$, when the mean is based on six observations, i.e.

$$P(\bar{x} > \mu + 2\cdot01 S(\bar{x})) = 0\cdot05$$

which means that a sample mean based on six observations will only exceed the true mean by more than 2·01 standard errors about 1 in 20 times.

We can also compute 95 per cent confidence limits for the true mean from

$$\bar{x} \pm t_{n-1}\,(P = 0.025)S(\bar{x})$$

or 99 per cent confidence limits from

$$\bar{x} \pm t_{n-1}\,(P = 0.005)S(\bar{x})$$

Example

Suppose six determinations of some quantity have been carried out, the results being as follows:

$$4.71, \quad 4.69, \quad 4.73, \quad 4.70, \quad 4.68, \quad 4.74$$

We can use such data to estimate the true value of the measured quantity together with confidence limits, if we assume that there is no constant error present. Alternatively, when the true value is known, perhaps from theoretical considerations or when it is some generally accepted standard that is being measured, we can estimate the bias in the technique from such data. We shall use the above data to illustrate both cases.

Since estimates of precision are based on the relative variability of the measurements, removal of a constant from all the measurements in order to simplify the arithmetic will not affect the standard deviation or the standard error. We shall for convenience subtract 4·68 from all the measurements, although 4·69 or 4·70 would do equally well. The six measurements on x simplify by the transformation $y = x - 4.68$ to:

$$0.03, \quad 0.01, \quad 0.05, \quad 0.02, \quad 0.00, \quad 0.06$$

$$n = 6$$

$$\sum y = 0.17, \qquad \bar{y} = \frac{\sum y}{n} = \frac{0.17}{6} = 0.0283$$

$$\sum y^2 = 0.0075, \qquad \frac{(\sum y)^2}{n} = \frac{(0.17)^2}{6} = \frac{0.0289}{6} = 0.0004817$$

The sample variance

$$S^2(y) = \frac{\sum y^2 - (\sum y)^2/n}{n-1} = \frac{0.0075 - 0.004817}{5}$$

$$= \frac{0.002683}{5} = 0.0005366$$

$$= S^2(x)$$

The standard deviation of a single observation is

$$S(x) = \sqrt{(0.0005366)} = 0.0232$$

N

The standard error $S(\bar{x})$ is given by

$$S(\bar{x}) = \frac{S(x)}{\sqrt{n}} = \frac{0 \cdot 0232}{\sqrt{6}} = \frac{0 \cdot 0232}{2 \cdot 449}$$

$$= 0 \cdot 00947$$

$$\bar{x} = 4 \cdot 68 + \bar{y} = 4 \cdot 68 + 0 \cdot 0283 = 4 \cdot 7083$$

The 95 per cent confidence limits for the true mean are given by

$$\bar{x} \pm t_{n-1}\,(P = 0 \cdot 025)S(\bar{x}),$$

i.e. $4 \cdot 7083 \pm t_5\,(P = 0 \cdot 025) \times 0 \cdot 00947$

$$= 4 \cdot 7083 \pm 2 \cdot 57 \times 0 \cdot 00947$$

$$= 4 \cdot 7083 \pm 0 \cdot 0243$$

$$= 4 \cdot 6840 \quad \text{and} \quad 4 \cdot 7326$$

The true value is therefore estimated as $4 \cdot 708$ with 95 per cent confidence limits $4 \cdot 684$ and $4 \cdot 733$.

If in this example we knew that the true value of the measurement was $4 \cdot 65$ we could use the data to estimate the bias in the technique.

We are given $\mu = 4 \cdot 65$.

The bias is $\bar{x} - \mu = 4 \cdot 708 - 4 \cdot 65 = +0 \cdot 058$.

The 95 per cent confidence limits for the bias are given by

$$(\bar{x} - \mu) \pm t_{n-1}\,(P = 0 \cdot 025)S(\bar{x})$$

i.e. $0 \cdot 058 \pm 0 \cdot 024$

$$= 0 \cdot 034 \quad \text{and} \quad 0 \cdot 082$$

Thus there is evidence of a positive bias in the technique of amount $0 \cdot 058$ with 95 per cent confidence limits $0 \cdot 034$ and $0 \cdot 082$.

There is no evidence of bias if the confidence limits contain the value zero in their range.

We shall now consider the situation when two sets of observations on some quantity are available. The number of observations in each set need not be equal. The results may be denoted by

$$x_1, x_2, \ldots, x_m$$

$$y_1, y_2, \ldots, y_n$$

where we have m observations in the first set and n observations in the second set. The results might have arisen from the measurement of the same quantity by two different methods. There are two questions which we need to answer:

(i) Is the error the same in both sets of data?

(ii) Do both sets of observations have the same mean?

The first step is to calculate the mean and the variance for both sets separately. Thus for

1st set: $\qquad \bar{x} = \dfrac{\sum x}{m}, \qquad S^2(x) = \dfrac{\sum x^2 - (\sum x)^2/m}{m-1},$

2nd set: $\qquad \bar{y} = \dfrac{\sum y}{n}, \qquad S^2(y) = \dfrac{\sum y^2 - (\sum y)^2/n}{n-1}$

To test whether the errors are the same for both sets of data we take the ratio of the larger estimate of variance to the smaller. The ratio S^2_{max}/S^2_{min} has an F-distribution with v_1 and v_2 degrees of freedom. We write

$$F_{v_1, v_2} = \frac{\text{Max } S^2}{\text{Min } S^2}$$

where $v_1 =$ number of degrees of freedom for max S^2
and $\quad v_2 =$ number of degrees of freedom for min S^2.

Some critical values of F are given in Table 3, based on the 5 per cent probability level:

TABLE 3

$v_2 \backslash v_1$	1	2	3	4	5	6	10
1	648	800	864	900	922	937	969
2	38·5	39·0	39·2	39·2	39·3	39·3	39·4
3	17·4	16·0	15·4	15·1	14·9	14·7	14·4
4	12·2	10·7	9·98	9·60	9·36	9·20	8·84
5	10·0	8·43	7·76	7·39	7·15	6·98	6·62
6	8·81	7·26	6·60	6·23	5·99	5·82	5·46

When the computed F-ratio exceeds the corresponding tabulated value we may assert that the two variances are significantly different at the 5 per cent probability level. We shall only consider the comparison of two means when both sets of data are equally precise, that is when the computed F-ratio is not significant. In this case we have two independent estimates of the same common variance σ^2; $S^2(x)$ based on $(m-1)$ degrees of freedom and $S^2(y)$ based on $(n-1)$ degrees of freedom. Both estimates of variance can be combined to give the pooled estimate of variance

$$S^2 = \frac{(m-1)S^2(x) + (n-1)S^2(y)}{m+n-2}$$

$$= \frac{\sum x^2 - (\sum x)^2/m + \sum y^2 - (\sum y)^2/n}{m+n-2}$$

The standard error of the difference between the two experimental means is given by

$$S(\bar{x}-\bar{y}) = \sqrt{\left[S^2\left(\frac{1}{m}+\frac{1}{n}\right)\right]}$$

The t-test for the comparison of the two means $\bar{x}$ and $\bar{y}$ is based on $(m+n-2)$ degrees of freedom and merely expresses the observed difference $\bar{x}-\bar{y}$ in terms of its standard error, i.e.

$$t_{m+n-2} = \left|\frac{\bar{x}-\bar{y}}{s(\bar{x}-\bar{y})}\right| = \left|\frac{\bar{x}-\bar{y}}{\sqrt{\left[S^2\left(\frac{1}{m}+\frac{1}{n}\right)\right]}}\right|$$

If the computed value of t_{m+n-2} ($P = 0.025$) exceeds the tabulated value of t_{m+n-2} ($P = 0.025$) then there is evidence that the two means are different at the 5 per cent probability level. An example is given below to illustrate the steps in the computation.

Example

Some quantity has been determined by two different methods. The results are given below:

Method 1: 30·01, 30·00, 30·03, 30·04, 30·01, 30·03
Method 2: 30·07, 30·04, 30·06, 30·05, 30·08.

Is there any evidence that the two methods are not equally precise or that the means of the two methods are not the same?

To simplify the arithmetic we shall subtract 30·00 from all the observations before starting the computation.
The transformations are:

$$u = x - 30.00$$

and

$$v = y - 30.00$$

Method 1

$$m = 6$$

$$\sum u = 0.12$$

$$\bar{u} = 0.02$$

$$\sum u^2 = 0.0036$$

$$\frac{(\sum u)^2}{m} = \frac{(0.12)^2}{6} = 0.0024$$

$$\sum (u-\bar{u})^2 = \sum u^2 - \frac{(\sum u)^2}{m}$$
$$= 0.0036 - 0.0024$$
$$= 0.0012$$

$$S^2(u) = \frac{0.0012}{6-1} = 0.00024$$
$$= S^2(x)$$

Method 2

$n = 5$

$\sum v = 0 \cdot 30$

$\bar{v} = 0 \cdot 06$

$\sum v^2 = 0 \cdot 0190$

$\dfrac{(\sum v)^2}{n} = \dfrac{(0 \cdot 30)^2}{5} = 0 \cdot 0180$

$$\sum (v - \bar{v})^2 = \sum v^2 - \dfrac{(\sum v)^2}{n}$$

$$= 0 \cdot 0190 - 0 \cdot 0180$$

$$= 0 \cdot 0010$$

$$S^2(v) = \dfrac{0 \cdot 0010}{5 - 1} = 0 \cdot 00025$$

$$= S^2(y)$$

The larger estimate of variance is that for method 2, based on 4 degrees of freedom. It is therefore $S^2(y)$ that is divided by $S^2(x)$ to obtain the F-ratio:

$$F_{4,5} = \frac{S^2(y)}{S^2(x)} = \frac{0 \cdot 00025}{0 \cdot 00024} = 1 \cdot 04$$

Since $F_{4,5}$ is less than $F_{4,5}$ $(P = 0 \cdot 05) = 7 \cdot 39$, there is no evidence that the two methods have different precisions.

The pooled estimate of variance

$$S^2 = \frac{0 \cdot 0012 + 0 \cdot 0010}{6 + 5 - 2} = \frac{0 \cdot 0022}{9}$$

$$= 0 \cdot 0002444$$

The standard error of the difference between the means of the two methods is given by

$$S(\bar{x} - \bar{y}) = \sqrt{\left[S^2 \left(\frac{1}{m} + \frac{1}{n} \right) \right]} = \sqrt{\left[0 \cdot 0002444 \left(\frac{1}{6} + \frac{1}{5} \right) \right]}$$

$$= \sqrt{(0 \cdot 000089628)}$$

$$= 0 \cdot 00947$$

To test the significance of the difference between $\bar{x}$ and $\bar{y}$ we compute

$$t_{m+n-2} = \left| \frac{\bar{x} - \bar{y}}{S(\bar{x} - \bar{y})} \right|$$

$$t_{6+5-2} = \left| \frac{30 \cdot 02 - 30 \cdot 06}{0 \cdot 00947} \right|$$

$$t_9 = \frac{0 \cdot 040}{0 \cdot 00947} = 4 \cdot 22 > t_9 \ (P = 0 \cdot 025) = 2 \cdot 26$$

Since t_9 exceeds the tabulated value of t_9 $(P = 0\cdot025) = 2\cdot26$, $\bar{x}$ and $\bar{y}$ are significantly different.

The mean value for method 2 is $0\cdot040$ higher than method 1 with 95 per cent confidence limits

$$0\cdot040 \pm t_9 \, (P = 0\cdot025) S(\bar{x} - \bar{y})$$

$$= 0\cdot040 \pm 2\cdot26 \times 0\cdot00947$$

$$= 0\cdot040 \pm 0\cdot021$$

$$= 0\cdot019 \quad \text{and} \quad 0\cdot061$$

Linear Regression

In many experiments the value of a random variable y is observed at n different values of some independent variable x. The n pairs of observations may be denoted by

$$(x_1, y_1), (x_2, y_2), \ldots, (x_i, y_i), \ldots, (x_n, y_n)$$

It is often necessary to obtain a relationship between x and y, when one exists. The first step should always be to plot the n points on graph paper. From the graph it is generally apparent whether the two variables are related or not, and when related whether the relationship is an approximately linear one or not. When the relationship between x and y is not linear it is frequently possible to represent the relationship to within the experimental uncertainty by a polynomial equation

$$y = a_0 + a_1 x + a_2 x^2 + \ldots + a_p x^p$$

where p is some low integer. The error in using

$$a_0 + a_1 x_i + a_2 x_i^2 + \ldots + a_p x_i^p$$

to predict y_i is

$$y_i - (a_0 + a_1 x_i + a_2 x_i^2 + \ldots + a_p x_i^p) = e_i \quad \text{(say)}$$

The constants $a_0, a_1, \ldots, a_p$ in the polynomial of degree p of "best fit" are selected in such a way as to minimize the sum of the squares of the residuals e_i, that is, by minimizing

$$S = \sum_{i=1}^{n} e_i^2 = \sum_{i=1}^{n} (y_i - a_0 - a_1 x_i - a_2 x_i - \ldots - a_p x^p)^2$$

When S is a minimum

$$\frac{\partial S}{\partial a_0} = 0, \quad \frac{\partial S}{\partial a_1} = 0, \ldots, \quad \frac{\partial S}{\partial a_p} = 0$$

These $(p + 1)$ equations in the $(p + 1)$ unknowns may be solved as a set of simultaneous linear equations to give values for $a_0, a_1, \ldots, a_p$.

The above method of estimation is an instance of the method of Least Squares.

The above method will be illustrated for the simple case when x and y are linearly related by the equation

$$y = a_0 + a_1 x$$

In this case

$$S = \sum_{i=1}^{n} (y_i - a_0 - a_1 x_i)^2$$

$$\frac{\partial S}{\partial a_0} = \sum_{i=1}^{n} -2(y_i - a_0 - a_1 x_i)$$

and

$$\frac{\partial S}{\partial a_1} = \sum_{i=1}^{n} -2x_i(y_i - a_0 - a_1 x_i)$$

On equating each partial derivative to zero, dividing both sides by -2 and rearranging, we get

$$na_0 + \sum_{i=1}^{n} x_i a_1 = \sum_{i=1}^{n} y_i,$$

$$\sum_{i=1}^{n} x_i a_0 + \sum_{i=1}^{n} x_i^2 a_1 = \sum_{i=1}^{n} x_i y_i$$

It is necessary to calculate

$$\sum_{i=1}^{n} x_i, \quad \sum_{i=1}^{n} y_i, \quad \sum_{i=1}^{n} x_i^2 \quad \text{and} \quad \sum_{i=1}^{n} x_i y_i$$

from the data. These values can be inserted in the two equations and a_0 and a_1 determined. The solutions of these equations may be expressed in the form

$$a_1 = \frac{\sum_{i=1}^{n} x_i y_i - \sum_{i=1}^{n} x_i \sum_{i=1}^{n} y_i/n}{\sum_{i=1}^{n} x_i^2 - \left(\sum_{i=1}^{n} x_i\right)^2/n}$$

and

$$a_0 = \bar{y} - a_1 \bar{x}.$$

The values of a_0 and a_1 are inserted in the equation $y = a_0 + a_1 x$, to obtain the line of "best fit" for estimating y in terms of x. This equation is known as the linear regression equation of y on x. a_1 is referred to as the linear regression coefficient of y on x and is the best estimate of the slope of the line; it measures the change in y per unit change in x.

When the x measurements are error free, and the ordinates y are normally distributed about the line with equal variance σ^2, for each value of x we can estimate σ^2 by

$$S^2 = \frac{\left(\sum y^2 - \dfrac{(\sum y)^2}{n}\right) - \dfrac{[\sum xy - (\sum x \sum y)/n]^2}{\sum x^2 - (\sum x)^2/n}}{n-2}$$

Using S^2, we can attach 95 per cent confidence limits to the estimated gradient a_1, by computing

$$a_1 \pm t_{n-2}(P = 0\cdot025)\sqrt{\left(\frac{S^2}{\sum x^2 - (\sum x)^2/n}\right)}$$

a_0 measures the intercept of the line on the y-axis, that is, it is the initial response. Ninety-five per cent confidence limits may be attached to a_0 by computing

$$a_0 \pm t_{n-2}(P = 0\cdot025)\sqrt{\left[S^2\left(\frac{1}{n} + \frac{\bar{x}^2}{\sum x^2 - (\sum x)^2/n}\right)\right]}$$

When the line of "best fit" is used to estimate y for some value x_0, say, of x; by substituting $x = x_0$ in the equation $y = a_0 + a_1 x$ we get $y = a_0 + a_1 x_0$. The 95 per cent confidence limits on this estimate are obtained from

$$(a_0 + a_1 x_0) \pm t_{n-2}(P = 0\cdot025)\sqrt{\left[S^2\left(\frac{1}{n} + \frac{(x_0 - \bar{x})^2}{\sum x^2 - (\sum x)^2/n}\right)\right]}.$$

We shall illustrate the computation required in fitting a straight line and in obtaining confidence limits on the various quantities that have been estimated, by means of an example.

Example

The following five pairs of observations are available, from which we require to estimate the linear equation for predicting y in terms of x:

x	0·04	0·36	0·41	0·42	0·44
y	117	96	90	88	86

The first step is to plot the points to get a visual impression of the type of relationship between x and y, as shown in Fig. 3.

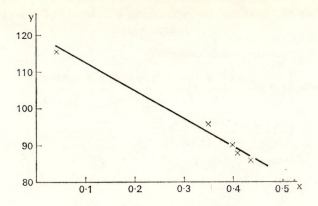

FIG. 3. Linear regression of y on x.

The relationship between x and y is approximately linear. To estimate the equation $y = a_0 + a_1 x$, we compute

$$n = 5$$

$$\sum x = 1{\cdot}67 \qquad\qquad \sum y = 477$$

$$\sum x^2 = 0{\cdot}6693 \qquad\qquad \sum xy = 150{\cdot}94$$

$$\frac{(\sum x)^2}{n} = 0{\cdot}55778 \qquad\qquad \frac{\sum x \sum y}{n} = 159{\cdot}318$$

$$\sum x^2 - \frac{(\sum x)^2}{n} = 0{\cdot}11152 \qquad \sum xy - \frac{\sum x \sum y}{n} = -8{\cdot}378$$

$$\sum y^2 = 46145$$

$$\frac{(\sum y)^2}{n} = 45505{\cdot}8$$

$$\sum y^2 - \frac{(\sum y)^2}{n} = 639{\cdot}2$$

$$\therefore \quad a_1 = \frac{\sum xy - \sum x \sum y / n}{\sum x^2 - (\sum x)^2 / n} = \frac{-8{\cdot}378}{0{\cdot}11152}$$

$$= 75{\cdot}16$$

and

$$a_0 = \bar{y} - a_1 \bar{x}$$

$$= \frac{477}{5} - (-75{\cdot}16)\frac{1{\cdot}67}{5}$$

$$= 95{\cdot}4 + (75{\cdot}16 \times 0{\cdot}334)$$

$$= 120{\cdot}5$$

N*

Thus the line of "best fit" for estimating y in terms of x is given by
$$y = 120 \cdot 5 - 75 \cdot 16x$$
The residual variance is estimated by

$$S^2 = \left[\left(\sum y^2 - \frac{(\sum y)^2}{n}\right) - \frac{(\sum xy - \sum x \sum y/n)^2}{\sum x^2 - (\sum x)^2/n}\right] \Big/ (n-2)$$

$$= \left(639 \cdot 2 - \frac{(-8 \cdot 378)^2}{0 \cdot 11152}\right) \Big/ 3$$

$$= (639 \cdot 2 - 629 \cdot 6)/3$$

$$= \frac{9 \cdot 6}{3}$$

$$= 3 \cdot 2$$

Ninety-five per cent confidence limits on the estimated gradient are given by:

$$a_1 \pm t_{n-2}(P = 0 \cdot 025) \Big/ \left(\sqrt{\frac{S^2}{\sum x^2 - (\sum x)^2/n}}\right)$$

$$= -75 \cdot 16 \pm t_3(P = 0 \cdot 025) \Big/ \left(\sqrt{\frac{3 \cdot 2}{0 \cdot 11152}}\right)$$

$$= -75 \cdot 16 \pm 3 \cdot 18 \sqrt{(28 \cdot 6944)}$$

$$= -75 \cdot 16 \pm (3 \cdot 18 \times 5 \cdot 36)$$

$$= -75 \cdot 16 \pm 17 \cdot 04$$

$$= -92 \cdot 2 \quad \text{and} \quad -58 \cdot 1$$

The 95 per cent confidence limits on the "initial response" a_0 are given by

$$a_0 \pm t_3(P = 0 \cdot 025) \Big/ \left[\sqrt{S^2\left(\frac{1}{n} + \frac{\bar{x}^2}{\sum x^2 - (\sum x)^2/n}\right)}\right]$$

$$= 120 \cdot 5 \pm 3 \cdot 18 \Big/ \left[\sqrt{3 \cdot 2\left(\frac{1}{5} + \frac{(0 \cdot 334)^2}{0 \cdot 11152}\right)}\right]$$

$$= 120 \cdot 5 \pm 3 \cdot 18 \sqrt{(3 \cdot 2 \times 1 \cdot 200)}$$

$$= 120 \cdot 5 \pm (3 \cdot 18 \times 1 \cdot 96) = 120 \cdot 5 \pm 6 \cdot 2$$

$$= 114 \cdot 3 \quad \text{and} \quad 126 \cdot 7$$

If we require to use the above equation to estimate y when $x = 0 \cdot 20$, the estimate is obtained by substituting $x = 0 \cdot 20$ in the equation
$$y = 120 \cdot 5 - 75 \cdot 16x$$

$$\therefore \quad y = 120 \cdot 5 - (75 \cdot 16 \times 0 \cdot 20)$$

$$= 105 \cdot 4$$

The 95 per cent confidence limits on this estimate are given by

$$105\cdot4 \pm t_3(P = 0\cdot025)\sqrt{\left[S^2\left(\frac{1}{n} + \frac{(x_0 - \bar{x})^2}{\sum x^2 - (\sum x)^2/n}\right)\right]}$$

$$= 105\cdot4 \pm 3\cdot18\sqrt{\left[3\cdot2\left(\frac{1}{5} + \frac{(0\cdot20 - 0\cdot334)^2}{0\cdot11152}\right)\right]}$$

$$= 105\cdot4 \pm 3\cdot18\sqrt{(3\cdot2 \times 0\cdot3165)}$$

$$= 105\cdot4 \pm 3\cdot18\sqrt{(1\cdot0128)}$$

$$= 105\cdot4 \pm 3\cdot20$$

$$= 102\cdot2 \quad \text{and} \quad 108\cdot6$$

Maximum Errors

Frequently, because of time limitations, it is not possible to repeat measurements in the physical chemistry laboratory. In such situations the lack of replication precludes the determination of the precision of the experimentally determined quantities. However, from knowledge of the maximum errors in directly measured quantities such as weights, volumes, pressures, temperatures, etc., maximum errors in results derived from them may be calculated.

A temperature, read by eye, from a thermometer graduated in tenths of a degree Centigrade may be in error by at most $\pm 0\cdot1°C$, if we discount constant errors. A burette calibrated correctly every $0\cdot1$ ml may be read, using a lens, to within $\pm 0\cdot02$ ml. In this way maximum errors may be placed on all direct measurements.

Most experimental results of interest are obtained by computing some function of directly measured quantities $x, y, z \ldots$. Thus if we denote the indirect quantity by u,

$$u = u(x, y, z, \ldots)$$

For example, the molecular weight M is obtained from measurements of pressure, volume, temperature and weight using the formula

$$M = \frac{wRT}{PV}$$

where R is the gas constant. A method is therefore required for estimating the maximum error $|\Delta M|$ in M, when maximum errors $|\Delta w|$, $|\Delta T|$, $|\Delta P|$ and $|\Delta V|$ are known.

Denoting the maximum positive errors in $x, y, z \ldots$ by $|\Delta x|$, $|\Delta y|$, $|\Delta z| \ldots$, and the consequent maximum positive error in u by

$|\Delta u|$, it follows from Taylor's expansion by neglecting squares and higher powers of $|\Delta x|$, $|\Delta y|$, $|\Delta z|$, . . .

$$|\Delta u| \simeq \left|\frac{\partial u}{\partial x}\right| |\Delta x| + \left|\frac{\partial u}{\partial y}\right| |\Delta y| + \left|\frac{\partial u}{\partial z}\right| |\Delta z| + \cdots \tag{1}$$

This approximation is reasonably good for errors which are only a few per cent of the true measurements.

The sum or difference of two measurements is very frequently used; for example, the difference of two temperatures is used for determining the rise in temperature. Generalizing,

$$u = x \pm y$$

$$\left|\frac{\partial u}{\partial x}\right| = 1 \quad \text{and} \quad \left|\frac{\partial u}{\partial y}\right| = |\pm 1| = 1$$

Therefore from (1)

$$|\Delta u| = |\Delta x| + |\Delta y|$$

Example

$$\text{Initial temperature,} \quad \theta_0 = 25° \pm 0{\cdot}1°\text{C.}$$
$$\text{Final temperature,} \quad \theta_1 = 75° \pm 0{\cdot}1°\text{C.}$$
$$\therefore \quad \text{Rise in temp.} \quad \theta_1 - \theta_0 = 50° \pm 0{\cdot}2°\text{C.}$$

If $u = xyz$

$$\frac{\partial u}{\partial x} = yz \quad \frac{\partial u}{\partial y} = xz \quad \frac{\partial u}{\partial z} = xy$$

$$\therefore \quad |\Delta u| = |yz| |\Delta x| + |xz| |\Delta y| + |xy| |\Delta z|$$

which on division by $u = uyz$, reduces to

$$\frac{|\Delta u|}{|u|} = \frac{|\Delta x|}{|x|} + \frac{|\Delta y|}{|y|} + \frac{|\Delta z|}{|z|}$$

Multiplication by 100 gives the relationship between the percentage errors, namely,

Percentage error in a product = Sum of the percentage errors in the individual measurements

Example

In a calorimetric determination, the heat of neutralization is obtained from the product of three quantities, namely,

$$\text{Heat of neutralization} = \frac{\text{Quantity}}{\text{taken}} \times \frac{\text{Total water}}{\text{equivalent}} \times \text{Temp. rise}$$

If the percentage errors in the three quantities on the right-hand side of the above equation are 1, 2, and 4 per cent, respectively, the percentage error in the heat of neutralization is the sum of these three percentages, that is ± 8 per cent.

The following important results are given without proof:

$$\% \text{ error in } x^n = n \text{ times } \% \text{ error in } x,$$

$$\% \text{ error in } x/y = \text{sum of } \% \text{ errors in } x \text{ and } y.$$

$$\text{Error in } \log_e x = \frac{|\Delta x|}{|x|} = \text{relative error in } x.$$

Example

For a first-order reaction,

$$kt = \log_e \frac{a}{a-c}$$

If, in an experiment, $a/(a-c) = 2$, the per cent errors in a and $a-c$ are both 2 per cent and $t = 10 \cdot 0 \pm 0 \cdot 1$ minutes; we can apply the above results to determine the maximum error in the reaction-rate constant k.

$$k = \frac{1}{t} \log_e \frac{a}{a-c} = \frac{1}{10} \log_e 2 = 0 \cdot 0693$$

Observing that k is the quotient of $\log_e a/(a-c)$ and t we may apply the above result for the percentage error in a quotient.

$$\% \text{ error in } k = \% \text{ error in } t + \% \text{ error in } \log_e \frac{a}{a-c}$$

$$= \frac{|\Delta t|}{t} \times 100 + \frac{\text{Error in } \log_e a/(a-c) \times 100}{\log_e a/(a-c)}$$

$$= \frac{0 \cdot 1 \times 100}{10} + \frac{\% \text{ error in } a + \% \text{ error in } (a-c)}{\log_e 2}$$

$$= 1 + \frac{2+2}{0 \cdot 693} = 1 + \frac{4}{0 \cdot 693}$$

$$\simeq 7\%$$

$$|\Delta k| \simeq \frac{7}{100} k = \frac{7}{100} \times 0 \cdot 0694 = 0 \cdot 005$$

$$\therefore \quad k = 0 \cdot 069 \pm 0 \cdot 005 \, \text{min}^{-1}$$

The same result may be obtained using the general formula (1), as follows:

$$k = \frac{1}{t}\log_e\frac{a}{a-c} = \frac{1}{t}\log_e\frac{a}{x}$$

where $a - c = x$.

$$\left|\frac{\partial k}{\partial t}\right| = \left|-\frac{1}{t^2}\log_e\frac{a}{a-c}\right| = \frac{k}{t}$$

$$\left|\frac{\partial k}{\partial a}\right| = \frac{1}{ta} \qquad \left|\frac{\partial k}{\partial x}\right| = \left|\frac{-1}{tx}\right| = \frac{1}{tx}$$

$$\therefore \quad |\Delta k| = \left|\frac{\partial k}{\partial t}\right||\Delta t| + \left|\frac{\partial k}{\partial a}\right||\Delta a| + \left|\frac{\partial k}{\partial x}\right||\Delta x|$$

$$= k\frac{|\Delta t|}{t} + \frac{1}{t}\frac{|\Delta a|}{a} + \frac{1}{t}\frac{|\Delta x|}{x}$$

$$= 0{\cdot}0693 \times \frac{0{\cdot}1}{10} + \frac{1}{10}\frac{2}{100} + \frac{1}{10}\frac{2}{100}$$

$$= 0{\cdot}004693 \simeq 0{\cdot}005\,\text{min}^{-1}$$

List of Textbooks

Theoretical

G. M. Barrow, *Physical Chemistry*, McGraw-Hill, New York (1966).

S. Glasstone, *Textbook of Physical Chemistry*, Macmillan, London (1948).

S. H. Maron and C. F. Prutton, *Principles of Physical Chemistry*, Macmillan, London (1965).

E. A. Moelwyn-Hughes, *Physical Chemistry*, Pergamon, Oxford (1962).

W. J. Moore, *Physical Chemistry*, Longmans, London (1964).

J. R. Partington, *An Advanced Treatise on Physical Chemistry*, Longmans, London (1949).

Thermodynamics

S. Glasstone, *Thermodynamics for Chemists*, Macmillan, London (1955).

J. G. Aston and J. J. Fritz, *Thermodynamics and Statistical Thermodynamics*, John Wiley, New York (1959).

Phase Equilibria

S. T. Bowen, *The Phase Rule and Phase Reactions*, Macmillan, London (1954).

Light and Spectra

C. N. Banwell, *Fundamentals of Molecular Spectroscopy*, McGraw-Hill, New York (1966).

G. M. Barrow, *Introduction to Molecular Spectroscopy*, McGraw-Hill, New York (1962).

Chemical Kinetics

K. J. Laidler, *Reaction Kinetics*, Vols. 1 and 2, Pergamon Press.

A. Frost and R. Pearson, *Kinetics and Mechanism*, John Wiley, New York (1962).

Surface Chemistry and Colloids

A. W. Adamson, *Physical Chemistry of Surfaces*, Interscience, New York (1960).

D. J. Shaw, *Introduction to Colloid and Surface Chemistry*, Butterworths, London (1966).

Electrochemistry

A. R. Denaro, *Elementary Electrochemistry*, Butterworths, London (1965).

R. A. Robinson and R. H. Stokes, *Electrolyte Solutions*, Butterworths, London (1959).

E. J. King, *Acid–Base Equilibria, The International Encyclopedia of Physical Chemistry and Chemical Physics*, edited by E. A. Guggenheim, J. E. Mayer and F. C. Tompkins, Pergamon, Oxford (1965).

Practical

H. D. Crockford and J. W. Nowell, *Laboratory Manual of Physical Chemistry*, John Wiley, New York (1956).

F. Daniels, J. H. Matthews, J. W. Williams, P. Bender and R. A. Alberty, *Experimental Physical Chemistry*, McGraw-Hill, New York (1956).

A. W. Davison, H. S. van Klooster, W. H. Bauer and G. J. Janz, *Laboratory Manual of Physical Chemistry*, John Wiley, New York (1956).

A. Findlay, *Practical Physical Chemistry* (revised by J. A. Kitchener), Longmans, London (1959).

A. M. JAMES, *Practical Physical Chemistry*, J. and A. Churchill, London (1967).

R. LIVINGSTON, *Physico Chemical Experiments*, Macmillan, New York (1957).

W. G. PALMER, *Experimental Physical Chemistry*, Cambridge University Press (1941).

D. P. SHOEMAKER and C. W. GARLAND, *Experiments in Physical Chemistry*, McGraw-Hill Book Co., New York (1962).

A. WEISSBERGER (Ed.), *Physical Methods of Organic Chemistry*, vol. 1, pts. I and II, Interscience Publishers (1949).

C. D. HODGMAN (Ed.), *Handbook of Chemistry and Physics*, Chemical Rubber Publishing Co.

List of Instruments and Manufacturers

THIS appendix contains a list of instruments which are being used in the various student experiments in our laboratories. It is not suggested that these are the only instruments which could be used but they are listed here as we have found them to be satisfactory for those experiments in which they are employed.

Calorimeter
Bomb calorimeter and accessories.
Baird and Tatlock Ltd., Chadwell Heath, Essex.
Exps. 16, 96.

Conductance and capacitance bridges with accessories
(1) Conductivity bridge with visual indicator.
Cambridge Instrument Co. Ltd., 13 Grosvenor Place, London, S.W.1.
Exps. 39, 43.
(2) Transistorized conductivity bridge.
Doran Instrument Co. Ltd., Stroud, Glos.
Exp. 37, 38.
(3) Conductivity measuring bridge (Philips)
Research and Control Instruments Ltd., 207 King's Cross Road, London, W.C.1.
Exp. 30.
(4) Wayne Kerr universal bridge. B221
The Wayne Kerr Laboratories Ltd., Roebuck Road, Chessington, Surrey.
Exps. 42, 53, 76, 92, 98.

Du Noüy tensiometer
Cambridge Instrument Co. Ltd., 13 Grosvenor Place, London, S.W.1.
Exp. 111.

Electrophoresis apparatus (*continuous*)
Shandon Scientific Co. Ltd., 6 Cromwell Place, London, S.W.7.
Exp. 75.

Galvanometers
Spot galvanometer.
Cambridge Instrument Co. Ltd., 13 Grosvenor Place, London, S.W.1.
Exps. 44, 45, 46, 47, 51, 60, 80, 84, 85, 98, 115.

Gas chromatograph
Argon chromatograph.
W. G. Pye and Co. Ltd., Granta Works, Cambridge.
Exps. 56, 57, 90.

Organic quenched standard liquid counting tube. Type M6
20th Century Electronics Ltd., Centronics Works, King Henry's Drive, New Addington, Croydon, Surrey.
Exps. 88, 89.

Geiger–Müller Counting equipment. Type 1221
Ericsson Telephones Ltd., Instrument Division, High Church Street, New
Basford, Nottingham.
Exps. 88, 89.

pH meters
(1) Cambridge bench pH meter.
Cambridge Instrument Co. Ltd., 13 Grosvenor Place, London, S.W.1.
Exps. 41, 48, 49, 50, 79, 87.
(2) Universal pH meter and millivoltmeter.
Exps. 68, 69.
(3) Potentiometric pH meter (11088),
Exp. 81.
W. G. Pye and Co. Ltd., Granta Works, Cambridge.

Polarimeters. Model A
Bellingham and Stanley Ltd., 71 Hornsey Rise, London, N.19.
Exps. 28, 70, 109.

Polarographs
(1) Manual polarograph.
Cambridge Instrument Co. Ltd., 13 Grosvenor Place, London, S.W.1.
Exp. 85.
(2) Pen recording polarograph.
Cambridge Instrument Co. Ltd., 13 Grosvenor Place, London, S.W.1.
Exps. 86, 87, 102, 114.

Potentiometers
(1) Cambridge slide wire potentiometer.
Cambridge Instrument Co. Ltd., 13 Grosvenor Place, London, S.W.1.
Exps. 60, 80.
(2) Muirhead D.C. potentiometer (D–972–A).
Muirhead and Co. Ltd., Beckenham, Kent.
Exps. 44, 45, 46, 84.
(3) Tinsley potentiometer
H. Tinsley and Co. Ltd., Werndee Hall, South Norwood, London, S.E.25.
Exps. 47, 51, 78, 115.

Power packs
(1) Dual stabilized D.C. supply. Type PP3.
Advance Components Ltd., Roebuck Road, Hainault, Ilford, Essex.
Exps. 13, 40, 68.
(2) Shandon stabilized D.C. power supply. Type 2523 Mk. II.
Shandon Scientific Co. Ltd., 6 Cromwell Place, London, S.W.7.
Exp. 77.

Refractometer
Abbé refractometer.
Gallenkamp and Co. Ltd., Technico House, Sun Street, London, E.C.2.
Exps. 3, 18, 53, 54, 92.

Spectrometer
  NMR Spectrometer (R 10).
  Exp. 105.
  Perkin-Elmer Ltd., Beaconsfield, Bucks., England.

Spectrophotometers
  Unicam SP.200 infra-red spectrophotometer.
  Exps. 65, 66, 102.
  Unicam SP.500 ultraviolet and visible spectrophotometer.
  Exps. 71, 81, 100, 112.
  Unicam SP.600.
  Exps. 25, 26, 64, 79, 108.
  Unicam SP.800.
  Exps. 81, 100, 101.
  Unicam Instruments, Arbury Works, Cambridge.
  Uvispeck Spectrophotometer.
  Exps. 81, 100, 101, 112.
  Hilger and Watts, 98 St. Pancras Way, London, N.W.1.

Spectrograph
  Hilger medium quartz spectrograph.
  Hilger and Watts, 98 St. Pancras Way, London, N.W.1.
  Exp. 63.

LOGARITHMS OF NUMBERS
AND ANTI-LOGARITHMS

LOGARITHMS

	0	1	2	3	4	5	6	7	8	9	1	2	3	4	5	6	7	8	9
10	0000	0043	0086	0128	0170						4	9	13	17	22	26	30	34	38
						0212	0253	0294	0334	0374	4	8	12	16	20	24	28	32	36
11	0414	0453	0492	0531	0569						4	8	12	16	20	23	27	31	35
						0607	0645	0682	0719	0755	4	8	12	15	19	22	26	30	33
12	0792	0828	0864	0899	0934						3	7	11	14	18	21	25	28	32
						0969	1004	1038	1072	1106	3	7	10	14	17	21	24	27	31
13	1139	1173	1206	1239	1271						3	7	10	13	17	20	23	26	30
						1303	1335	1367	1399	1430	3	7	10	13	16	19	22	26	29
14	1461	1492	1523	1553	1584						3	6	9	13	16	19	22	25	28
						1614	1644	1673	1703	1732	3	6	9	12	15	18	21	24	26
15	1761	1790	1818	1847	1875						3	6	9	11	14	17	20	23	26
						1903	1931	1959	1987	2014	3	6	9	11	14	17	20	23	25
16	2041	2068	2095	2122	2148						3	6	8	11	14	16	19	22	25
						2175	2201	2227	2253	2279	3	5	8	10	13	16	18	21	24
17	2304	2330	2355	2380	2405						3	5	8	10	13	16	18	21	23
						2430	2455	2480	2504	2529	3	5	8	10	12	15	17	20	22
18	2553	2577	2601	2625	2648						2	5	7	10	12	14	17	19	21
						2672	2695	2718	2742	2765	2	4	7	9	11	14	16	18	21
19	2788	2810	2833	2856	2878						2	4	6	9	11	13	16	18	20
						2900	2923	2945	2967	2989	2	4	6	8	11	13	15	17	19
20	3010	3032	3054	3075	3096	3118	3139	3160	3181	3201	2	4	6	8	11	13	15	17	19
21	3222	3243	3263	3284	3304	3324	3345	3365	3385	3404	2	4	6	8	10	12	14	16	18
22	3424	3444	3464	3483	3502	3522	3541	3560	3579	3598	2	4	6	8	10	12	14	16	18
23	3617	3636	3655	3674	3692	3711	3729	3747	3766	3784	2	4	6	7	9	11	13	15	17
24	3802	3820	3838	3856	3874	3892	3909	3927	3945	3962	2	4	5	7	9	11	13	14	16
25	3979	3997	4014	4031	4048	4065	4082	4099	4116	4133	2	3	5	7	9	10	12	14	15
26	4150	4166	4183	4200	4216	4232	4249	4265	4281	4298	2	3	5	7	8	10	11	13	15
27	4314	4330	4346	4362	4378	4393	4409	4425	4440	4456	2	3	5	6	8	9	11	13	14
28	4472	4487	4502	4518	4533	4548	4564	4579	4594	4609	2	3	5	6	8	9	10	12	13
29	4624	4639	4654	4669	4683	4698	4713	4728	4742	4757	1	3	4	6	7	9	10	12	13
30	4771	4786	4800	4814	4829	4843	4857	4871	4886	4900	1	3	4	6	7	9	10	11	13
31	4914	4928	4942	4955	4969	4983	4997	5011	5024	5038	1	3	4	6	7	8	10	11	12
32	5051	5065	5079	5092	5105	5119	5132	5145	5159	5172	1	3	4	5	7	8	9	11	12
33	5185	5198	5211	5224	5237	5250	5263	5276	5289	5302	1	3	4	5	6	8	9	10	12
34	5315	5328	5340	5353	5366	5378	5391	5403	5416	5428	1	3	4	5	6	8	9	10	11
35	5441	5453	5465	5478	5490	5502	5514	5527	5539	5551	1	2	4	5	6	7	9	10	11
36	5563	5575	5587	5599	5611	5623	5635	5647	5658	5670	1	2	4	5	6	7	8	10	11
37	5682	5694	5705	5717	5729	5740	5752	5763	5775	5786	1	2	3	5	6	7	8	9	10
38	5798	5809	5821	5832	5843	5855	5866	5877	5888	5899	1	2	3	5	6	7	8	9	10
39	5911	5922	5933	5944	5955	5966	5977	5988	5999	6010	1	2	3	4	5	7	8	9	10
40	6021	6031	6042	6053	6064	6075	6085	6096	6107	6117	1	2	3	4	5	6	7	9	10
41	6128	6138	6149	6159	6170	6180	6191	6201	6212	6222	1	2	3	4	5	6	7	8	9
42	6232	6243	6253	6263	6274	6284	6294	6304	6314	6325	1	2	3	4	5	6	7	8	9
43	6335	6345	6355	6365	6375	6385	6395	6405	6415	6425	1	2	3	4	5	6	7	8	9
44	6435	6444	6454	6464	6474	6484	6493	6503	6513	6522	1	2	3	4	5	6	7	8	9
45	6532	6542	6551	6561	6571	6580	6590	6599	6609	6618	1	2	3	4	5	6	7	8	9
46	6628	6637	6646	6656	6665	6675	6684	6693	6702	6712	1	2	3	4	5	6	7	7	8
47	6721	6730	6739	6749	6758	6767	6776	6785	6794	6803	1	2	3	4	5	5	6	7	8
48	6812	6821	6830	6839	6848	6857	6866	6875	6884	6893	1	2	3	4	4	5	6	7	8
49	6902	6911	6920	6928	6937	6946	6955	6964	6972	6981	1	2	3	4	4	5	6	7	8
	0	1	2	3	4	5	6	7	8	9	1	2	3	4	5	6	7	8	9

	0	1	2	3	4	5	6	7	8	9	1	2	3	4	5	6	7	8	9
50	6990	6998	7007	7016	7024	7033	7042	7050	7059	7067	1	2	3	3	4	5	6	7	8
51	7076	7084	7093	7101	7110	7118	7126	7135	7143	7152	1	2	3	3	4	5	6	7	8
52	7160	7168	7177	7185	7193	7202	7210	7218	7226	7235	1	2	2	3	4	5	6	7	7
53	7243	7251	7259	7267	7275	7284	7292	7300	7308	7316	1	2	2	3	4	5	6	6	7
54	7324	7332	7340	7348	7356	7364	7372	7380	7388	7396	1	2	2	3	4	5	6	6	7
55	7404	7412	7419	7427	7435	7443	7451	7459	7466	7474	1	2	2	3	4	5	5	6	7
56	7482	7490	7497	7505	7513	7520	7528	7536	7543	7551	1	2	2	3	4	5	5	6	7
57	7559	7566	7574	7582	7589	7597	7604	7612	7619	7627	1	2	2	3	4	5	5	6	7
58	7634	7642	7649	7657	7664	7672	7679	7686	7694	7701	1	1	2	3	4	4	5	6	7
59	7709	7716	7723	7731	7738	7745	7752	7760	7767	7774	1	1	2	3	4	4	5	6	7
60	7782	7789	7796	7803	7810	7818	7825	7832	7839	7846	1	1	2	3	4	4	5	6	6
61	7853	7860	7868	7875	7882	7889	7896	7903	7910	7917	1	1	2	3	4	4	5	6	6
62	7924	7931	7938	7945	7952	7959	7966	7973	7980	7987	1	1	2	3	3	4	5	6	6
63	7993	8000	8007	8014	8021	8028	8035	8041	8048	8055	1	1	2	3	3	4	5	5	6
64	8062	8069	8075	8082	8089	8096	8102	8109	8116	8122	1	1	2	3	3	4	5	5	6
65	8129	8136	8142	8149	8156	8162	8169	8176	8182	8189	1	1	2	3	3	4	5	5	6
66	8196	8202	8209	8215	8222	8228	8235	8241	8248	8254	1	1	2	3	3	4	5	5	6
67	8261	8267	8274	8280	8287	8293	8299	8306	8312	8319	1	1	2	3	3	4	5	5	6
68	8325	8331	8338	8344	8351	8357	8363	8370	8376	8382	1	1	2	3	3	4	4	5	6
69	8388	8395	8401	8407	8414	8420	8426	8432	8439	8445	1	1	2	3	3	4	4	5	6
70	8451	8457	8463	8470	8476	8482	8488	8494	8500	8506	1	1	2	2	3	4	4	5	6
71	8513	8519	8525	8531	8537	8543	8549	8555	8561	8567	1	1	2	2	3	4	4	5	5
72	8573	8579	8585	8591	8597	8603	8609	8615	8621	8627	1	1	2	2	3	4	4	5	5
73	8633	8639	8645	8651	8657	8663	8669	8675	8681	8686	1	1	2	2	3	4	4	5	5
74	8692	8698	8704	8710	8716	8722	8727	8733	8739	8745	1	1	2	2	3	4	4	5	5
75	8751	8756	8762	8768	8774	8779	8785	8791	8797	8802	1	1	2	2	3	3	4	5	5
76	8808	8814	8820	8825	8831	8837	8842	8848	8854	8859	1	1	2	2	3	3	4	5	5
77	8865	8871	8876	8882	8887	8893	8899	8904	8910	8915	1	1	2	2	3	3	4	4	5
78	8921	8927	8932	8938	8943	8949	8954	8960	8965	8971	1	1	2	2	3	3	4	4	5
79	8976	8982	8987	8993	8998	9004	9009	9015	9020	9025	1	1	2	2	3	3	4	4	5
80	9031	9036	9042	9047	9053	9058	9063	9069	9074	9079	1	1	2	2	3	3	4	4	5
81	9085	9090	9096	9101	9106	9112	9117	9122	9128	9133	1	1	2	2	3	3	4	4	5
82	9138	9143	9149	9154	9160	9165	9170	9175	9180	9186	1	1	2	2	3	3	4	4	5
83	9191	9196	9201	9206	9212	9217	9222	9227	9232	9238	1	1	2	2	3	3	4	4	5
84	9243	9248	9253	9258	9263	9269	9274	9279	9284	9289	1	1	2	2	3	3	4	4	5
85	9294	9299	9304	9309	9315	9320	9325	9330	9335	9340	1	1	2	2	3	3	4	4	5
86	9345	9350	9355	9360	9365	9370	9375	9380	9385	9390	1	1	2	2	3	3	4	4	5
87	9395	9400	9405	9410	9415	9420	9425	9430	9435	9440	0	1	1	2	2	3	3	4	4
88	9445	9450	9455	9460	9465	9469	9474	9479	9484	9489	0	1	1	2	2	3	3	4	4
89	9494	9499	9504	9509	9513	9518	9523	9528	9533	9538	0	1	1	2	2	3	3	4	4
90	9542	9547	9552	9557	9562	9566	9571	9576	9581	9586	0	1	1	2	2	3	3	4	4
91	9590	9595	9600	9605	9609	9614	9619	9624	9628	9633	0	1	1	2	2	3	3	4	4
92	9638	9643	9647	9652	9657	9661	9666	9671	9675	9680	0	1	1	2	2	3	3	4	4
93	9685	9689	9694	9699	9703	9708	9713	9717	9722	9727	0	1	1	2	2	3	3	4	4
94	9731	9736	9741	9745	9750	9754	9759	9763	9768	9773	0	1	1	2	2	3	3	4	4
95	9777	9782	9786	9791	9795	9800	9805	9809	9814	9818	0	1	1	2	2	3	3	4	4
96	9823	9827	9832	9836	9841	9845	9850	9854	9859	9863	0	1	1	2	2	3	3	4	4
97	9868	9872	9877	9881	9886	9890	9894	9899	9903	9908	0	1	1	2	2	3	3	4	4
98	9912	9917	9921	9926	9930	9934	9939	9943	9948	9952	0	1	1	2	2	3	3	4	4
99	9956	9961	9965	9969	9974	9978	9983	9987	9991	9996	0	1	1	2	2	3	3	3	4
	0	1	2	3	4	5	6	7	8	9	1	2	3	4	5	6	7	8	9

	0	1	2	3	4	5	6	7	8	9	1	2	3	4	5	6	7	8	9
·00	1000	1002	1005	1007	1009	1012	1014	1016	1019	1021	0	0	1	1	1	1	2	2	2
·01	1023	1026	1028	1030	1033	1035	1038	1040	1042	1045	0	0	1	1	1	1	2	2	2
·02	1047	1050	1052	1054	1057	1059	1062	1064	1067	1069	0	0	1	1	1	1	2	2	2
·03	1072	1074	1076	1079	1081	1084	1086	1089	1091	1094	0	0	1	1	1	1	2	2	2
·04	1096	1099	1102	1104	1107	1109	1112	1114	1117	1119	0	1	1	1	1	2	2	2	2
·05	1122	1125	1127	1130	1132	1135	1138	1140	1143	1146	0	1	1	1	1	2	2	2	2
·06	1148	1151	1153	1156	1159	1161	1164	1167	1169	1172	0	1	1	1	1	2	2	2	2
·07	1175	1178	1180	1183	1186	1189	1191	1194	1197	1199	0	1	1	1	1	2	2	2	2
·08	1202	1205	1208	1211	1213	1216	1219	1222	1225	1227	0	1	1	1	1	2	2	2	3
·09	1230	1233	1236	1239	1242	1245	1247	1250	1253	1256	0	1	1	1	1	2	2	2	3
·10	1259	1262	1265	1268	1271	1274	1276	1279	1282	1285	0	1	1	1	1	2	2	2	3
·11	1288	1291	1294	1297	1300	1303	1306	1309	1312	1315	0	1	1	1	1	2	2	2	3
·12	1318	1321	1324	1327	1330	1334	1337	1340	1343	1346	0	1	1	1	2	2	2	2	3
·13	1349	1352	1355	1358	1361	1365	1368	1371	1374	1377	0	1	1	1	2	2	2	3	3
·14	1380	1384	1387	1390	1393	1396	1400	1403	1406	1409	0	1	1	1	2	2	2	3	3
·15	1413	1416	1419	1422	1426	1429	1432	1435	1439	1442	0	1	1	1	2	2	2	3	3
·16	1445	1449	1452	1455	1459	1462	1466	1469	1472	1476	0	1	1	1	2	2	2	3	3
·17	1479	1483	1486	1489	1493	1496	1500	1503	1507	1510	0	1	1	1	2	2	2	3	3
·18	1514	1517	1521	1524	1528	1531	1535	1538	1542	1545	0	1	1	1	2	2	2	3	3
·19	1549	1552	1556	1560	1563	1567	1570	1574	1578	1581	0	1	1	1	2	2	3	3	3
·20	1585	1589	1592	1596	1600	1603	1607	1611	1614	1618	0	1	1	1	2	2	3	3	3
·21	1622	1626	1629	1633	1637	1641	1644	1648	1652	1656	0	1	1	2	2	2	3	3	3
·22	1660	1663	1667	1671	1675	1679	1683	1687	1690	1694	0	1	1	2	2	2	3	3	3
·23	1698	1702	1706	1710	1714	1718	1722	1726	1730	1734	0	1	1	2	2	2	3	3	4
·24	1738	1742	1746	1750	1754	1758	1762	1766	1770	1774	0	1	1	2	2	2	3	3	4
·25	1778	1782	1786	1791	1795	1799	1803	1807	1811	1816	0	1	1	2	2	2	3	3	4
·26	1820	1824	1828	1832	1837	1841	1845	1849	1854	1858	0	1	1	2	2	3	3	3	4
·27	1862	1866	1871	1875	1879	1884	1888	1892	1897	1901	0	1	1	2	2	3	3	4	4
·28	1905	1910	1914	1919	1923	1928	1932	1936	1941	1945	0	1	1	2	2	3	3	4	4
·29	1950	1954	1959	1963	1968	1972	1977	1982	1986	1991	0	1	1	2	2	3	3	4	4
·30	1995	2000	2004	2009	2014	2018	2023	2028	2032	2037	0	1	1	2	2	3	3	4	4
·31	2042	2046	2051	2056	2061	2065	2070	2075	2080	2084	0	1	1	2	2	3	3	4	4
·32	2089	2094	2099	2104	2109	2113	2118	2123	2128	2133	0	1	1	2	2	3	3	4	4
·33	2138	2143	2148	2153	2158	2163	2168	2173	2178	2183	0	1	1	2	2	3	3	4	4
·34	2188	2193	2198	2203	2208	2213	2218	2223	2228	2234	1	1	2	2	3	3	4	4	5
·35	2239	2244	2249	2254	2259	2265	2270	2275	2280	2286	1	1	2	2	3	3	4	4	5
·36	2291	2296	2301	2307	2312	2317	2323	2328	2333	2339	1	1	2	2	3	3	4	4	5
·37	2344	2350	2355	2360	2366	2371	2377	2382	2388	2393	1	1	2	2	3	3	4	4	5
·38	2399	2404	2410	2415	2421	2427	2432	2438	2443	2449	1	1	2	2	3	3	4	4	5
·39	2455	2460	2466	2472	2477	2483	2489	2495	2500	2506	1	1	2	2	3	3	4	5	5
·40	2512	2518	2523	2529	2535	2541	2547	2553	2559	2564	1	1	2	2	3	3	4	5	5
·41	2570	2576	2582	2588	2594	2600	2606	2612	2618	2624	1	1	2	2	3	4	4	5	5
·42	2630	2636	2642	2648	2655	2661	2667	2673	2679	2685	1	1	2	2	3	4	4	5	6
·43	2692	2698	2704	2710	2716	2723	2729	2735	2742	2748	1	1	2	2	3	4	4	5	6
·44	2754	2761	2767	2773	2780	2786	2792	2799	2805	2812	1	1	2	3	3	4	4	5	6
·45	2818	2825	2831	2838	2844	2851	2858	2864	2871	2877	1	1	2	3	3	4	5	5	6
·46	2884	2891	2897	2904	2911	2917	2924	2931	2938	2944	1	1	2	3	3	4	5	5	6
·47	2951	2958	2965	2972	2979	2985	2992	2999	3006	3013	1	1	2	3	3	4	5	5	6
·48	3020	3027	3034	3041	3048	3055	3062	3069	3076	3083	1	1	2	3	3	4	5	6	6
·49	3090	3097	3105	3112	3119	3126	3133	3141	3148	3155	1	1	2	3	4	4	5	6	6
	0	1	2	3	4	5	6	7	8	9	1	2	3	4	5	6	7	8	9

	0	1	2	3	4	5	6	7	8	9	1 2 3	4 5 6	7 8 9
·50	3162	3170	3177	3184	3192	3199	3206	3214	3221	3228	1 1 2	3 4 4	5 6 7
·51	3236	3243	3251	3258	3266	3273	3281	3289	3296	3304	1 2 2	3 4 5	5 6 7
·52	3311	3319	3327	3334	3342	3350	3357	3365	3373	3381	1 2 2	3 4 5	5 6 7
·53	3388	3396	3404	3412	3420	3428	3436	3443	3451	3459	1 2 2	3 4 5	6 6 7
·54	3467	3475	3483	3491	3499	3508	3516	3524	3532	3540	1 2 2	3 4 5	6 6 7
·55	3548	3556	3565	3573	3581	3589	3597	3606	3614	3622	1 2 2	3 4 5	6 7 7
·56	3631	3639	3648	3656	3664	3673	3681	3690	3698	3707	1 2 3	3 4 5	6 7 8
·57	3715	3724	3733	3741	3750	3758	3767	3776	3784	3793	1 2 3	3 4 5	6 7 8
·58	3802	3811	3819	3828	3837	3846	3855	3864	3873	3881	1 2 3	4 4 5	6 7 8
·59	3890	3899	3908	3917	3926	3936	3945	3954	3963	3972	1 2 3	4 5 5	6 7 8
·60	3981	3990	3999	4009	4018	4027	4036	4046	4055	4064	1 2 3	4 5 6	6 7 8
·61	4074	4083	4093	4102	4111	4121	4130	4140	4150	4159	1 2 3	4 5 6	7 8 9
·62	4169	4178	4188	4198	4207	4217	4227	4236	4246	4256	1 2 3	4 5 6	7 8 9
·63	4266	4276	4285	4295	4305	4315	4325	4335	4345	4355	1 2 3	4 5 6	7 8 9
·64	4365	4375	4385	4395	4406	4416	4426	4436	4446	4457	1 2 3	4 5 6	7 8 9
·65	4467	4477	4487	4498	4508	4519	4529	4539	4550	4560	1 2 3	4 5 6	7 8 9
·66	4571	4581	4592	4603	4613	4624	4634	4645	4656	4667	1 2 3	4 5 6	7 9 10
·67	4677	4688	4699	4710	4721	4732	4742	4753	4764	4775	1 2 3	4 5 7	8 9 10
·68	4786	4797	4808	4819	4831	4842	4853	4864	4875	4887	1 2 3	4 6 7	8 9 10
·69	4898	4909	4920	4932	4943	4955	4966	4977	4989	5000	1 2 3	5 6 7	8 9 10
·70	5012	5023	5035	5047	5058	5070	5082	5093	5105	5117	1 2 3	5 6 7	8 9 10
·71	5129	5140	5152	5164	5176	5188	5200	5212	5224	5236	1 2 4	5 6 7	8 10 11
·72	5248	5260	5272	5284	5297	5309	5321	5333	5346	5358	1 2 4	5 6 7	9 10 11
·73	5370	5383	5395	5408	5420	5433	5445	5458	5470	5483	1 2 4	5 6 7	9 10 11
·74	5495	5508	5521	5534	5546	5559	5572	5585	5598	5610	1 3 4	5 6 8	9 10 12
·75	5623	5636	5649	5662	5675	5689	5702	5715	5728	5741	1 3 4	5 7 8	9 10 12
·76	5754	5768	5781	5794	5808	5821	5834	5848	5861	5875	1 3 4	5 7 8	9 11 12
·77	5888	5902	5916	5929	5943	5957	5970	5984	5998	6012	1 3 4	5 7 8	10 11 12
·78	6026	6039	6053	6067	6081	6095	6109	6123	6138	6152	1 3 4	6 7 8	10 11 13
·79	6166	6180	6194	6209	6223	6237	6252	6266	6281	6295	1 3 4	6 7 9	10 11 13
·80	6310	6324	6339	6353	6368	6383	6397	6412	6427	6442	1 3 4	6 7 9	10 12 13
·81	6457	6471	6486	6501	6516	6531	6546	6561	6577	6592	1 3 4	6 7 9	11 12 14
·82	6607	6622	6637	6653	6668	6683	6699	6714	6730	6745	2 3 5	6 8 9	11 12 14
·83	6761	6776	6792	6808	6823	6839	6855	6871	6887	6902	2 3 5	6 8 9	11 13 14
·84	6918	6934	6950	6966	6982	6998	7015	7031	7047	7063	2 3 5	6 8 10	11 13 15
·85	7079	7096	7112	7129	7145	7161	7178	7194	7211	7228	2 3 5	7 8 10	11 13 15
·86	7244	7261	7278	7295	7311	7328	7345	7362	7379	7396	2 3 5	7 8 10	12 13 15
·87	7413	7430	7447	7464	7482	7499	7516	7534	7551	7568	2 3 5	7 9 10	12 14 16
·88	7586	7603	7621	7638	7656	7674	7691	7709	7727	7745	2 4 5	7 9 11	12 14 16
·89	7762	7780	7798	7816	7834	7852	7870	7889	7907	7925	2 4 5	7 9 11	13 14 16
·90	7943	7962	7980	7998	8017	8035	8054	8072	8091	8110	2 4 6	7 9 11	13 15 17
·91	8128	8147	8166	8185	8204	8222	8241	8260	8279	8299	2 4 6	8 9 11	13 15 17
·92	8318	8337	8356	8375	8395	8414	8433	8453	8472	8492	2 4 6	8 10 12	13 15 17
·93	8511	8531	8551	8570	8590	8610	8630	8650	8670	8690	2 4 6	8 10 12	14 16 18
·94	8710	8730	8750	8770	8790	8810	8831	8851	8872	8892	2 4 6	8 10 12	14 16 18
·95	8913	8933	8954	8974	8995	9016	9036	9057	9078	9099	2 4 6	8 10 12	15 17 19
·96	9120	9141	9162	9183	9204	9226	9247	9268	9290	9311	2 4 6	8 11 13	15 17 19
·97	9333	9354	9376	9397	9419	9441	9462	9484	9506	9528	2 4 6	9 11 13	15 17 20
·98	9550	9572	9594	9616	9638	9661	9683	9705	9727	9750	2 4 7	9 11 13	15 18 20
·99	9772	9795	9817	9840	9863	9886	9908	9931	9954	9977	2 5 7	9 11 14	16 18 20
	0	1	2	3	4	5	6	7	8	9	1 2 3	4 5 6	7 8 9

Index

The figures in **bold type** refer to experiments; the other figures refer to pages